Dependable Computing
and Fault-Tolerant Systems

Edited by
A. Avižienis, H. Kopetz, J. C. Laprie

Volume 2

U. Voges (ed.)

Software Diversity in Computerized Control Systems

Springer-Verlag Wien New York

Dipl.-Math. Udo Voges, Kernforschungszentrum Karlsruhe GmbH,
Karlsruhe, Federal Republic of Germany

With 41 Figures

Library of Congress Cataloging-in-Publication Data

Software diversity in computerized control systems.

   (Dependable computing and fault-tolerant systems ;
vol. 2)
   Contents: Introduction / U. Voges — Railway
applications / G. Haglin — Nuclear applications /
U. Voges, P. Bishop — [etc.]
   1. Fault-tolerant computing. 2. Computer software—
Reliability. 3. Automatic control—Data processing.
I. Voges, Udo, 1946—   . II. Series.
QA76.9.F38S65   1988   005   87-32367

ISBN-13:978-3-7091-8934-4 (U.S.)

ISSN 0932-5581
ISBN-13:978-3-7091-8934-4   e-ISBN-13:978-3-7091-8932-0
DOI: 10.1007/978-3-7091-8932-0

Preface

Software Diversity is one of the fault-tolerance means to achieve dependable systems.

In this volume, some experimental systems as well as real-life applications of software diversity are presented. The history, the current state-of-the-art and future perspectives are given.

Although this technique is used quite successfully in industrial applications, further research is necessary to solve some open questions. We hope to report on new results and applications in another volume of this series within some years.

Acknowledgements

The idea of the workshop was put forward by the chairpersons of IFIP WG 10.4, J.-C. Laprie, J. F. Meyer and Y. Tohma, in January 1986, and the editor of this volume was asked to organize the workshop.

This volume was edited with the assistance of the editors of the series, A. Avižienis, H. Kopetz and J.-C. Laprie, who also had the function of reviewers.

Karlsruhe, October 1987   U. Voges, Editor

Table of Contents

1

Introduction

Dependable computing is an issue which was already of much concern before this term got accepted and more widely used [Laprie 1985]. Especially the aspects of safety and reliability made the application of fault tolerance techniques necessary, since complete fault avoidance was impossible to achieve in general applications.

As the causes for software failures are different from those of hardware, different fault tolerance techniques are necessary. Software diversity is one of them.

Software Diversity has many faces and many names. It probably started by simply naming it *fault-tolerant programming* [Elmendorf 1972] and later *redundant programming* [Avižienis 1975], but in order to put more emphasis on the difference of the solutions, also *distinct software* (e. g. [Fischler 1975]), *dissimilar software* (e. g. [Martin 1982]), and *dual programming* (e. g. [Ramamoorthy 1981]) were used. In the course of further application of this technique, *N-Version Programming* (NVP) (e. g. [Avižienis 1977]) and *Multi-Version-Software* (MVS) (e. g. [Kelly 1982]) were other terms which were used. Besides Software Diversity NVP and MVS are now the most often used terms.

Another technique which is very much related to NVP is the Recovery Block technique [Randell 1975]. The main difference between these two approaches is that the Recovery Block approach makes use of an acceptance test and activates the secondary alternate only in case of a detected error, while in NVP all versions are activated all the time, and the acceptance test is replaced by a comparison of the different outputs. For a more complete comparison see Table 1.

As an extension and generalization of this technique, the term *Design Diversity* is emerging, including not only Software Diversity, but also Hardware Diversity [Avižienis 1986].

The idea of applying diversity is not as new as it might be expected after

Table 1. Comparison Between N-Version Programming and Recovery Blocks

N-Version Programming	Recovery Blocks
error detection by comparison	error detection by acceptance testing
error correction by masking and voting	error correction by recovery and activation of secondary alternate
voting and majority decision (relative test)	acceptance testing (absolute test)
majority required	one accepted alternate required
always execution of all alternates	execution of n+1 alternates only if n are erroneous
parallel execution normal serial execution possible	serial execution normal parallel execution possible
static redundancy	dynamic redundancy

these remarks. In 1834, Dionysius Lardner writes ([Lardner 1961] p. 177):

> *The most certain and effectual check upon errors which arise in the process of computation, is to cause the same computations to be made by separate and independent computers; and this check is rendered still more decisive if they make their computations by different methods.*

But it was also realized that this is not the only solution and that this solution still contains drawbacks and so Lardner continues:

> *It is, nevertheless, a remarkable fact, that several computers, working separately and independently, do frequently commit precisely the same error; so that falsehood in this case assumes that character of consistency, which is regarded as the exclusive attribute of truth. Instances of this are familiar to most persons who have had the management of the computation of tables.*

But because of the application of computer systems in application areas with high dependability requirements the interest in software diversity has increased.

This volume contains as a main part written versions of presentations given on a one-and-a-half day Workshop on "Design Diversity in Action", which was organized by the IFIP Working Group 10.4 "Reliable Computing and Fault Tolerance" on June 27 and 28, 1986 in Baden/Vienna in Austria.

The aim of this workshop was to bring together the greatest diversity of people who use software diversity in industrial applications as well as those which are conducting experiments and evaluations. A list of questions was sent to the speakers of the workshop in order to have some of the main points covered from different perspectives.

Since not all experiments and applications of Software Diversity neither could be presented at the Workshop nor could be included with separate papers in this volume, an annotated bibliography is included. Some historical remarks as well as references to a wide scope of experiments and applications related to software diversity are mentioned.

In addition, a report on a workshop on "Reliability Modelling for Fault-Tolerant Software" is included in this volume. It contains a list of open research issues in this area.

It is hoped that in some future time a new workshop on this topic is organized to draw together further application areas of software diversity, to show new ways to use it and to inform on gathered experiences. The editor will be happy to receive any information on additional experiments and applications.

References

[Avižienis 1975]   A. Avižienis, "Fault-Tolerance and Fault-Intolerance: Complementary Approaches to Reliable Computing," in *Proc. Intern. Conf. on Reliable Software,* Los Angeles, CA, USA: 21-23 April 1975, pp. 458-464.

[Avižienis 1977]   A. Avižienis and L. Chen, "On the Implementation of N-Version Programming for Software Fault-Tolerance During Program Execution," in *Proc. Compsac77,* Chicago, IL, USA: November 1977, pp. 149-155.

[Avižienis 1986]   A. Avižienis and J.-C. Laprie, "Dependable Computing: From Concepts to Design Diversity," *IEEE Proceedings,* Vol. 74, No. 5, May 1986, pp. 629-638.

[Elmendorf 1972]   W. R. Elmendorf, "Fault-Tolerant Programming," in *Proc. 2nd Intern. Symp. on Fault-Tolerant Computing FTCS'2,* Newton, MA, USA: 19-21 June 1972, pp. 79-83.

[Fischler 1975]   M. A. Fischler, O. Firschein, and D. L. Drew, "Distinct Software: An Approach to Reliable Computing," in *Proc. Second USA-Japan Computer Conference,* 1975, pp. 573-579.

[Kelly 1982]   J. P. J. Kelly, "Specification of Fault-Tolerant Multi-Version Software: Experimental Studies of a Design Diversity Approach," UCLA, Computer Science Department, Los Angeles, CA, USA, Tech. Rep. CSD-820927, September 1982.

[Laprie 1985]   J.-C. Laprie, "Dependable Computing and Fault Tolerance: Concepts and Terminology," in *Proc. 15th Intern. Symp. on Fault-Tolerant Computing FTCS'15,* Ann Arbor, MI, USA: 19-21 June 1985, pp. 2-11.

[Lardner 1961]   D. Lardner, "Babbage's Calculating Engine; From the Edinburgh Review, July, 1834, No. CXX," in *Charles Babbage and His Calculating Engines,* E. Morrison, Ed. Dover Publications, Inc. New York, 1961.

[Martin 1982]   D. J. Martin, "Dissimilar Software in High Integrity Applications in Flight Controls," in *Proc. AGARD Symp. on Software Avionics, CPP-330,* The Hague, The Netherlands: September 1982, pp. 36.1-36.13.

[Ramamoorthy 1981]   C. V. Ramamoorthy, Y. R. Mok, F. B. Bastani, G. H. Chin, and K. Suzuki, "Application of a Methodology for the Development and Validation of Reliable Process Control Software," *IEEE Trans. on Software Engineering,* Vol. SE-7, No. 6, November 1981, pp. 537-555.

[Randell 1975]   B. Randell, "System Structure for Software Fault Tolerance," *IEEE Trans. on Software Eng.,* Vol. SE-1, No. 2, June 1975, pp. 220-232.

2

Railway Applications

The application of Software Diversity in industrial applications beyond an experimental stage was apparently done first in a railway system. This first system as well as newer developments are described in the following paper by Hagelin.

Bengt Sterner from the Statens Jarnvagar (Swedish State Railways) was one who encouraged the use of diversity in the Gothenburg system. His involvement in this project can also be seen by the formal specification language used for the interlocking system [Sterner 1978], which was named STERNOL.

In an Italian train control system functionally diverse programs are running in sequence, and their results are compared for error detection [Frullini 1984].

For the Singapore Mass Rapid Transit Railway, part of the system is realized with two diverse redundant microprocessors having different design objectives and diverse software in order to minimize the number of common mode failures [Davies 1984].

For licensing purposes for a railway safety system in Germany, not a parallel development, but an independent reverse development - from program object code to requirement specification - was performed with final comparison of the two documents, original requirements specification and the back translated one. This is another kind of application of the diversity principle [Krebs 1984].

The General Railway Signal Co. of Rochester, NY, USA, designed its computerized interlocking system called Vital Processor Interlocking with a single Intel8086 processor running two diverse programs written in assembly language. In case of a failure, a second back-up system (standby spare) takes over. The system is in operation in two railway companies in USA [Turner 1987].

A similar design is used by the Union Switch & Signal Co. of Pittsburgh,

PA, USA [Turner 1987].

There exist some further applications of Software Diversity in the railway environment, which, however, have not left the experimental stage, e. g. systems for a subway train control in Germany [Kapp 1981] and also in France.

Some experimentation was made in Germany by Siemens and the federal railway systems (DB) with the use of diversity, but up to now the general opinion has been to rely on identical hardware and software, using simple redundancy means and applying a rather strict verification and validation procedure [Schwier 1987].

References

[Davies 1984]   P. A. Davies, "The Latest Developments in Automatic Train Control," in *Proc. Intern. Conf. on Railway Safety Control and Automation Towards the 21st Century,* London, UK: 25-27 September 1984, pp. 272-279.

[Frullini 1984]   R. Frullini and A. Lazzari, "Use of Microprocessor in Fail-Safe on Board Equipment," in *Proc. Intern. Conf. on Railway Safety Control and Automation Towards the 21st Century,* London, UK: 25-27 September 1984, pp. 292-299.

[Kapp 1981]   K.-H. Kapp, R. Daum, E. Sartori, and R. Harms, "Sicherheit durch vollständige Diversität (Safety through complete Diversity - in German)," in *Proc. Fachtagung Prozeßrechner 1981,* München, FRG: Springer-Verlag Berlin-Heidelberg-New York, 10-11 March 1981, pp. 216-229.

[Krebs 1984]   H. Krebs and U. Haspel, "Ein Verfahren zur Software-Verifikation (A Technique for Software Verification - in German)," *Regelungstechnische Praxis,* Vol. 26, 1984, pp. 73-78.

[Schwier 1987]   W. Schwier, *Private communication,* 1987.

[Sterner 1978]   B. J. Sterner, "Computerised Interlocking System - a Multidimensional Structure in the Pursuit of Safety," *IMechE Railway Engineer International,* November/December 1978, pp. 29-30.

[Turner 1987]   D. B. Turner, R. D. Burns, and H. Hecht, "Designing Micro-Based Systems for Fail-Safe Travel," *IEEE Spectrum,* Vol. 24, No. 2, February 1987, pp. 58-63.

ERICSSON Safety System
for Railway Control

Gunnar Hagelin
ERICSSON SIGNAL SYSTEMS AB
P.O. Box 42 505
S-126 12 Stockholm
Sweden

1. Introduction

Traditionally, equipment used by railways have been grouped in vital and non-vital equipment. Vital equipment shall work in a fail-safe way. Inter-lockings and level crossing units are examples of vital equipment, and the train dispatching panel is an example of non-vital equipment.

Fail-safe is in signalling industry defined as "a characteristic of a system which ensures that any malfunction affecting safety will cause the system to revert to a state that is generally known to be safe". A safe state is for instance a signal showing a stop aspect (red light).

In a very strict way, railway systems can be defined to be fault tolerant.

The fail-safe requirements have forced railway equipment designers to use special techniques and special components. You cannot use a standard (tele-phone) relay in fail-safe circuits because there is a risk that two contacts weld together in a way that cannot be detected. Therefore, special Safety Relays have been developed.

Special components however are more expensive than standard components and relays are nowadays more expensive than electronic components. Therefore there is a demand to put electronics and computers into vital equipment. But how do you handle the fail-safe requirement in computer hardware and software? It is quite impossible to completely avoid faults in complex software systems, especially in real-time systems. And even if the software is fault-free, how do you *prove* that? Therefore you must accept that there always remain faults in software systems. In vital systems the effects of these faults must be detected and made non-dangerous.

The ERICSSON solution of this problem is DIVERSITY. This means that we design two different packages of software. They are both executed in the process and the results are compared and must be equal. This principle is now used in computerized Interlockings and Automatic Train Control systems in use in several countries. In the following chapters I will give short descriptions of these systems and after that discuss the methods we use at ERICSSON to achieve safe systems.

2. Interlockings

Interlockings are used in railway stations. The purposes of interlockings are to

● safeguard the movements of the trains,

● prevent dangerous situations,

● handle operator commands,

● inform the operator(s),

● control and supervise wayside equipment, like signals and point machines.

Traditionally, interlockings have been designed with mechanical or electromechanical components. ERICSSON started to work with computerized interlockings in the middle of the 1970's. The first installation was put into commercial operation in 1978 in Gothenburg (Sweden).

In 1978 we still used relays in parts of the system to interface the signals and points. Now we have a second generation of the system in operation. This second generation is fully electronic.

The basic design of ERICSSON computer interlockings is shown in Fig. 1. In the signal box we have the Central Equipment with

● interlocking computer(s)

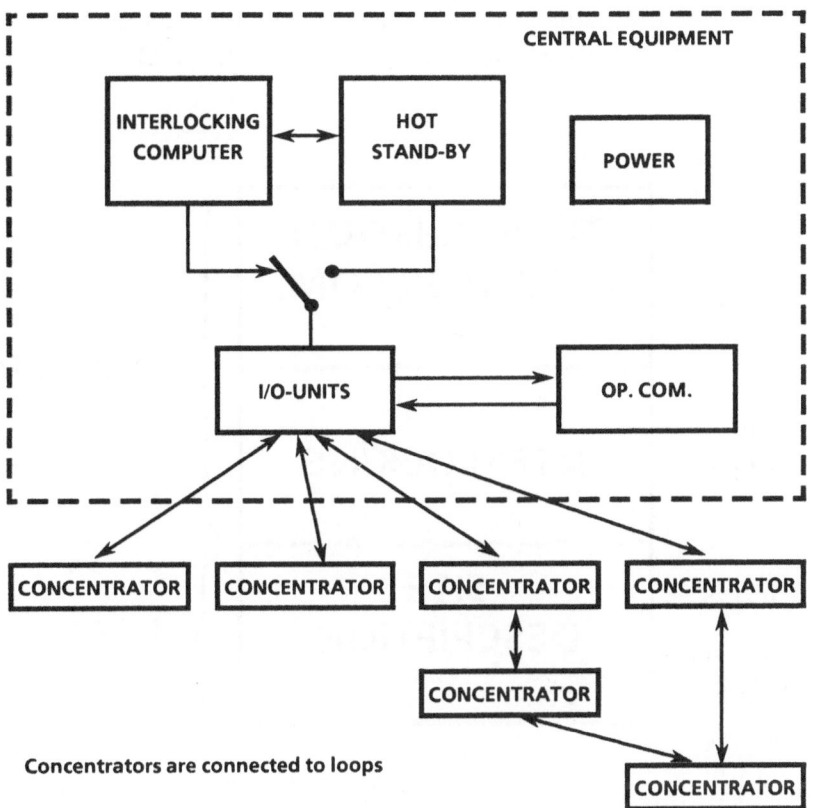

Fig. 1. Interlocking System Layout

- operator interface (VDU's and keyboards)
- power supplies.

Outdoors in the station area we have concentrators. Signals, point machines etc are connected to the concentrator. Between the concentrators and the central equipment we use a serial transmission system, organized in loops. The concentrators are located close to the controlled objects (to reduce cable lengths etc). In the concentrators we have

- transmission equipment and
- object controllers.

In the first generation we designed the object controllers with relays, in the second generation they are fully electronic.

The software in the interlocking computer can be divided into three parts (see Fig. 2). The programs to handle transmission, operator communication etc are written in assembler. These programs are "universal" in all installations.

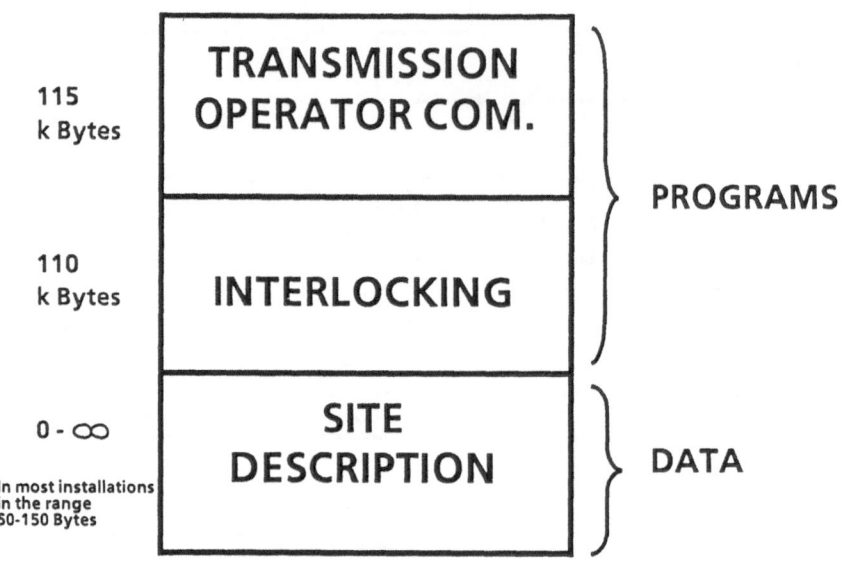

115
k Bytes

110
k Bytes

0 - ∞

In most installations
in the range
50-150 Bytes

TRANSMISSION
OPERATOR COM.

INTERLOCKING

SITE
DESCRIPTION

} PROGRAMS

} DATA

Fig. 2. Interlocking Memory Layout

The interlocking programs are handling the interlocking conditions in the stations. These programs are the same in all stations in each market. Because of different signalling requirements, these programs have to be adopted to each market (country, railway, company). The interlocking programs are written in a language called STERNOL, which is specially designed to handle signalling problems. STERNOL is a language, restricted to handle Boolean variables and some very simple arithmetic calculations. The programs are documented in a way which is similar to traditional relay diagrams (see Fig. 3).

In the object controllers we use electronics and microprocessors today. The software in these are written in Assembler and PL/M. Site data, which are the site description for the general programs, are produced by an off-line system. This is the part varying from installation to installation. All site dependent information is contained in this description.

```
-1    — U6=0 — K=3 —

 1    — K=4 ┬R2=1 ┬┬ I215=1 — U0<>2 — U0<>3 — R4=7      — R6=6      ┬TC=0 ┬
            ├R2=2 ┤                                                  └T2=0 ┤
            └R2=3 ┘└ I215=0 ┬U0=1 ┬┬TC=0 ─────────────────────────────────┤
                            └U0=0 ┘└U6=1 ┬ I205=1 — U0=1 ─┘
                                         └T2=0  — I205=0 ┘

 2    ┬K=4 ─────────────────────────┬┬ U0=2 ─────────────────┐
      └M<>3 — MM=0 ┬K=0 ┤└U0=3 ┬K=0 ─────────────┤
                   └K=5 ┘       ├TC=0 ─────────────┤
                                ├T2=0 — I205=0 ┤
                                └T0=0 — I005=0 ┘

 3    — K=4 — U0=3 — TC=1 ┬T2=1 ┬┬T0=1 ┬ ┬
                          └I205=1 ┘└I005=1 ┘

 4    — U0=4 — TC=1 — K=4 ┬T0=1 ┬┬T2=1 ┬ ┬
                          └I005=1 ┘└I205=1 ┘

 5    — R2=6 — K=4 — U2<>2 — U2<>3 ┬R6=1 ┬┬U0=5 ─────────────┐
                                   ├R6=2 ┤├U6=5 ─────────────┤
                                   └R6=3 ┘└U2=0 — U6=0 — PK=0 ┘

 6    — U0=6 ┬K=4 ─────────────────────────┐
            └M<>3 — MM=0 ┬K=0 ┬┬U2=2 ┤
                         └K=5 ┘└U2=3 ┘
```

Fig. 3. Examples of STERNOL statements

3. ATC

In Railway Technology ATC stands for Automatic Train Control. ATC shall

- help and supervise the drivers,
- warn the drivers and/or apply the brakes in all situations that can be dangerous.

ATC systems normally have three parts:

- wayside equipment to pick up information from signals, interlockings etc,
- on-board equipment to supervise the driving,
- transmission between wayside equipment and on-board equipment.

The basic design of the system is shown in Fig. 4. The information from the signals and interlockings is handled by electronic encoders. In the case of computer interlocking, data is fed directly to the beacons without the use of encoders. The beacons, which are powered from the passing trains, are sending the information to the locomotives. On board of the train microprocessors handle the information from the antenna, indicate it to the driver and process it together with information from the brake conduit and the speedometer.

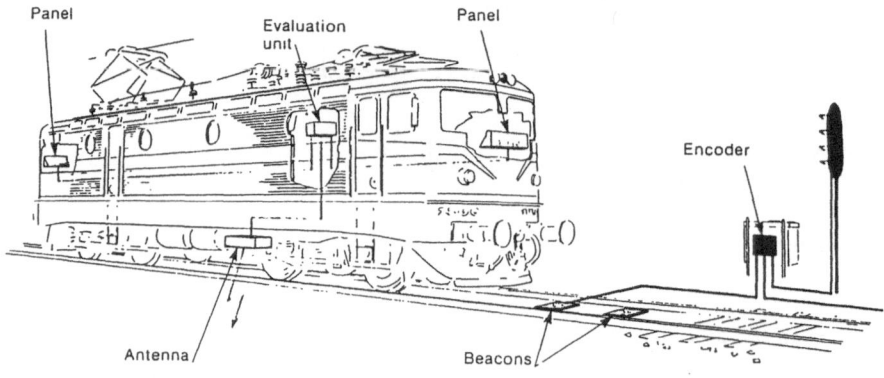

Fig. 4. ATC System Layout

The software in the ATC computer is written in Assembler and PL/M. The size of the software is about 32k byte.

ERICSSON modern ATC-systems now are in operation in many countries. The first was operating in Sweden starting 1978.

4. Methods

The basic design principle in ERICSSON Safety Systems is DIVERSITY. In software this means that vital functions are designed independently by two teams. Both systems are executed in the processors. The results of the calculations have to match before they are regarded to be correct.

Fig. 5 shows how diversity is used in the interlocking system. The two interlocking systems (A and B) are running in the interlocking computer. The comparators and detectors are built together into units (= object controllers) located in the concentrators. Each signal and point machine in the station has an individual object controller.

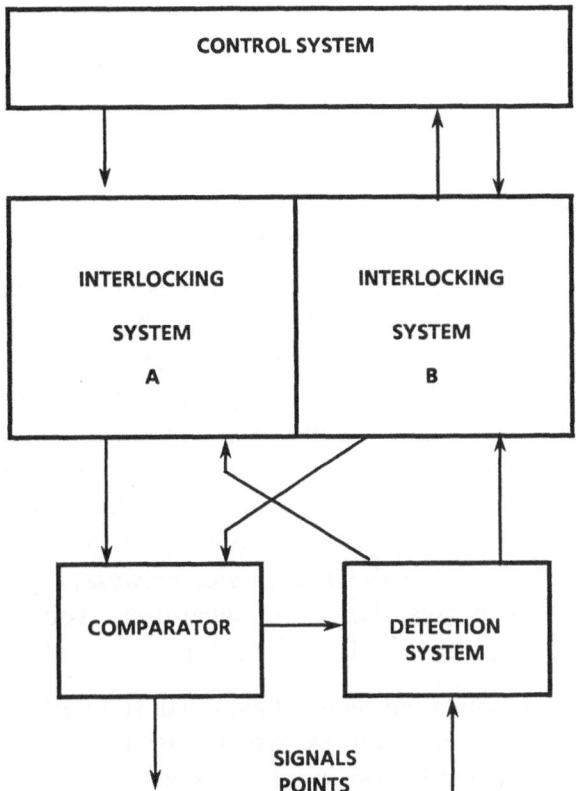

Fig. 5. Safety Layout of Computer Based Interlocking Systems JZS 750 and JZS 850

Diversity is achieved by

- terms and
- design rules.

Diversity is enforced by

- external coding of data and
- data organization.

External coding is achieved by coding the status messages from the object controllers. The Hamming distance between A- and B-messages is in most cases equal to 4. The two design teams are manned with different people. People are not allowed to move between the teams: in one of our projects they were even located in different cities. The teams are producing their own set of documentation.

However, DIVERSITY is not the only method used in ERICSSON Safety

Systems. Other methods are

● use of closed loops,

● time (age) checks,

● verification and validation,

● two redundant clocks, and

● HW-checks.

It is very important that safety systems are designed as closed loops. However you can never be sure that a loop really is closed. Therefore it is necessary that any break in the loop is detected and actions are taken towards "the safe directions". This is handled in the following way in the interlocking. A command to signals etc has to be recalculated and retransmitted cyclically. The object controllers "know" this and are expecting commands to be repeated. If the retransmission will stop, for any reason, the object controllers will put signals to stop as a fail-safe operation. To handle the problem with "too old data", all messages in the transmission network, and some data in the computers have time tags. These are supervised. Too old data is regarded as faulty and will cause signals to go to stop.

Safety validations are performed during all phases in development work (see Fig. 6). Validations are based on Hearings. We also use Fault Tree Analysis and Fault Effect Analysis. Fault Tree Analysis is based on the questions "what can be dangerous and what can cause danger". Fault Effect Analysis is based on the questions "what happens if ...". Fault Effect Analysis is used both for safety and reliability, but is mostly used for hardware.

We also want to point out that it is very important to have a stringent development method. In our systems diversity has been used to the greatest extend within software design. We have very few hardware units which are designed in a diversified way (just some I/O circuits).

5. Specifications

A common questions is "How were your products specified". The answer is that the specifications mostly were written in clear Swedish language (written text). Talking about specifications you must however realize that a specification can be divided into at least

● functional requirements and

● quality requirements (MTBF, probability of dangerous errors etc).

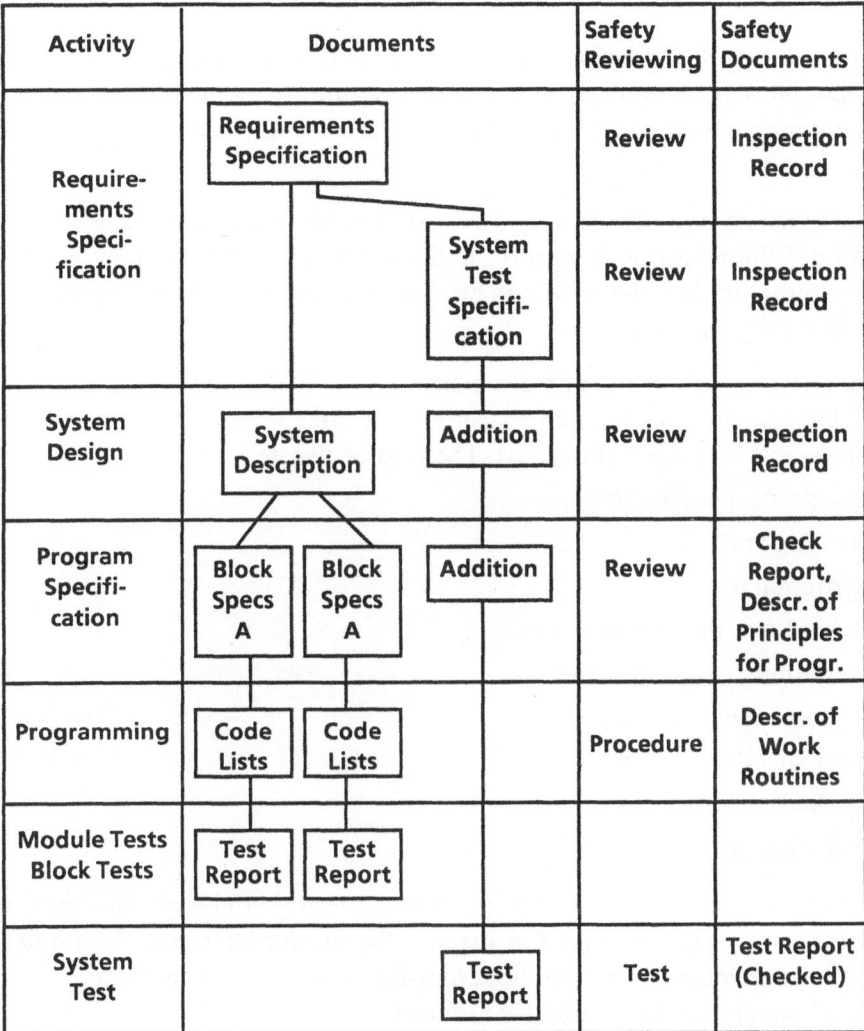

Activity	Documents			Safety Reviewing	Safety Documents
Require-ments Speci-fication	Requirements Specification			Review	Inspection Record
		System Test Specifi-cation		Review	Inspection Record
System Design	System Description	Addition		Review	Inspection Record
Program Specifi-cation	Block Specs A	Block Specs A	Addition	Review	Check Report, Descr. of Principles for Progr.
Programming	Code Lists	Code Lists		Procedure	Descr. of Work Routines
Module Tests Block Tests	Test Report	Test Report			
System Test			Test Report	Test	Test Report (Checked)

Fig. 6. Work Organization

The last point was (and still is) crucial when we started the work in the 1970's. Some customers said "New systems shall be as good or better than old systems". The problem is that nobody can give figures about the old systems.

6. Experience

Today, the systems described above are in operation in many installations in Sweden, Norway, Denmark and Finland. Within the next years they will be put into operation in some other countries like Switzerland, Turkey and Bulgaria.

ERICSSON experience is that diversity DOES pay. The answer to the question "Would we do it again?" is definitely YES. Up till now we have, after commissioning, detected one error that could have been dangerous in a Non-Diversity solution.

A common argument against Diversity is that it doubles the development cost. This is not true for two reasons. The first reason is that the system development is divided into steps. The cost of the steps

- program specification,
- coding and
- program test

is doubled, but the cost of the steps

- requirement specification,
- system specification,
- test specification and
- system test

is NOT doubled.

The second reason is that within a safety system you always find parts that have no safety requirements, for instance the printer interface. You must do some measurements to isolate these non-fail-safe from the fail-safe parts, but the work that has to be doubled is reduced.

7. Future

Our future work will go into two directions. The first is to introduce electronics and computers into fail-safe products that still are designed with relays. The second will be to introduce more modern techniques into our existing products. An example here is to use more high-level languages. We also must put more efforts into specification techniques.

References

[Andersson 1981]   H. Andersson and G. Hagelin: Computer Controlled Interlocking System, Ericsson Review, No. 2, 1981.

[Andersson 1983]   H. Andersson: Experience from the Introduction of ATC in Sweden, Ericsson Review, No. 1, 1983.

[Lind 1979]   O. Berg von Lind: Computers Can Now Perform Vital Functions Safely, Railway Gazette International, Nov. 1979.

[Nordenfors 1986]   D. Nordenfors and A. Sjöberg: Computer-Controlled Electronic Interlocking System. ERILOCK 850, Ericsson Review, No. 1, 1986.

[Sjöberg 1981]   A. Sjöberg: Automatic Train Control, Ericsson Review, No. 1, 1981.

3

Nuclear Applications

In the nuclear field, there have been several experiments with the application of software diversity. Three of them will be described in more detail in the following two papers. Voges reports on the results gathered in the early BPI-experiment and the design of the later MIRA-system, both done in Karlsruhe. Bishop then explains an experiment conducted by three countries. Besides these experiments, the following ones should be mentioned.

EPRI

In the late 1970's and early 1980's, under the direction of the Electric Power Research Institute (EPRI) of the USA an experiment was made which used diversity as one of different techniques to be applied in the project [Ramamoorthy 1981, Saib 1982]. The aim of the project was to develop a method for the construction of reliable software, which should be applied to a realistic nuclear power plant problem.

In this project participants came from The Babcock & Wilcox Company, Science Applications, Inc., University of California at Berkeley, and General Research Corporation.

Starting from the functional requirements, two teams developed independently a design, using the formal specification language RSL. After the verification of these designs, each team continued the development with detailed design, implementation, testing as well as validation and verification efforts. The final step was the comparison of the outputs of the two programs.

The main results of this project were:
- The use of automated tools can largely assist the development process.
- The savings in testing far outweigh the cost of dual program development.
- The methods applied improved the likelihood of detecting errors when introduced.

- The interface problems were reduced.
- The software correctness was increased compared to more conventional experience.

CANDU

For the Darlington Generation Station (4x850 MW), which is currently under construction, a computerized reactor safety shutdown system is developed. It consists of two components, SDS1 and SDS2. Each component has a similar computer configuration, but is from different manufacturers. Separate design teams develop the systems. Two programming languages are used: Fortran and Pascal. The complete system consists of 15 computers [Popovic 1986].

The previous design, which consisted of a different layout, is also using diverse hardware, but identical programming languages. So far no shutdown was due to a software error in those reactors which are running with a computerized shutdown system.

Halden

Within the Halden Reactor Project, a joint experiment was conducted between Halden (Norway) and the Technical Research Center VTT (Finland). Its aim was the evaluation of different software engineering methods. The application problem chosen was a reactor control system [Dahll 1979].

The specification was written in plain English. Two independent teams then started designing the system using different methods, one using structured flow diagrams, the other one using pseudo-code. The coding was done in Pascal and Fortran.

In order to check the efficiency of the testing procedures, error seeding technique was used in addition. The back-to-back test showed to be an efficient means of error detection.

Based on the experience in this experiment, a second experiment was designed, which is described in the paper by Bishop.

References

[Dahll 1979]   G. Dahll and J. Lahti, "An Investigation of Methods for Production and Verification of Highly Reliable Software," in *Proc. IFAC Workshop Safecomp'79*, Stuttgart, FRG: 16-18 May 1979, pp. 89-94.

[Popovic 1986]   J. R. Popovic, D. C. Chan, and D. B. Burjorjee, "Computer Control in Candu Plants," in *Symposium on Advanced Nuclear Services, CAN/CNS Intern. Nuclear Conference,* Toronto, CDN: 8-11 June 1986.

[Ramamoorthy 1981]   C. V. Ramamoorthy, Y. R. Mok, F. B. Bastani, G. H. Chin, and K. Suzuki, "Application of a Methodology for the Development and Validation of Reliable Process Control Software," *IEEE Trans. on Software Engineering,* Vol. SE-7, No. 6, November 1981, pp. 537-555.

[Saib 1982]   S. H. Saib, "Validation of Real-Time Software for Nuclear Plant Safety Applications," Tech. Rep. EPRI NP-2646, November 1982.

Use of Diversity
in Experimental Reactor Safety Systems

Udo Voges
Kernforschungszentrum Karlsruhe GmbH
Institut für Datenverarbeitung in der Technik
Postfach 3640, D-7500 Karlsruhe

Abstract

This paper describes two projects which were conducted at the Kernforschungszentrum Karlsruhe. The first was "BPI", a pilot implementation of parts of a reactor safety shut down system. In this experiment the problem was specified in natural language (German) with heavy use of mathematical notations. Based on this specification three teams prepared in parallel three implementations in three different languages.

The results of this experiment show that not only the errors made by the different teams were different, but also that the error detection capabilities were increased through the use of different teams. Therefore the overall reliability was higher than in a development environment without use of diversity.

The second project consisted of the design of the reactor safety shut down system "MIRA". Analogue to the triple modular redundant hardware structure of the system, three diverse versions of the application software should be installed. The design of the system as well as the reasons leading to the incorporation of software diversity are presented. It is anticipated that not only errors in those parts which are realized diversely can be tolerated to some extent, but also errors in those parts which are identical in the redundant system.

1. Introduction

In computer applications with high dependability requirements, like applications in safety-related areas, hardware as well as software has to have the ability to detect errors and to react to error situations. One possibility to increase the reliability of software is the use of diversity. Therefore some experiments were conducted at the Kernforschungszentrum Karlsruhe to evaluate the usability and the value of such an approach.

The first part of this paper reports on an experiment which took place between 1975 and 1979. After some introductory remarks on related work and definitions, the reactor protection system and the use of diversity within it are explained. Following this, the experiment and its results are described. Finally, possible achievements and experiences are summarized.

Not only newly developed reactors, but also existing power plants require the introduction of computerized systems. Maintenance of old systems is too costly or becomes almost impossible without introduction of modern technology which involves most often computers. If this change of technology takes place in safety critical parts, like safety shut-down systems, licensing authorities are also heavily involved. Within a research project at the Kernforschungszentrum Karlsruhe, a part of the hardwired core surveillance system should be replaced by a computerized system. In the second part of the paper the requirements for this system, the hardware structure chosen and the basic software layout will be described. Fault-avoidance techniques and fault-tolerance concepts including the diversity aspects which were planned will be explained. The experience gained in the first experiment was incorporated.

For hardware there exists already a broad knowledge of and much experience in the use of redundancy. Normally, hardware redundancy is realized by simply replicating identical components or even systems. The different components are operated in parallel and a majority voter decides on their results (m-out-of-n-systems with voter). Since the replicated components are developed according to the same specification and design logic, only aging errors and production-failures can be detected by this technique. If in addition logic design errors shall be detected, independent designs have to be made. This kind of redundancy is called diversity. E.g. in hardware, different components with identical functions are used or different physical effects are measured for the calculation of a certain quantity.

Simple software redundancy through duplication is normally of no help, since no error and fault tolerance capability is involved: the same errors are in the duplicated systems, and aging errors like in hardware do not occur. Therefore the need for diversity is even more important for software than for hardware.

2. Reactor Safety Systems and Software Diversity

Generally reactor safety systems are designed as hardware m-out-of-n-systems. This is also the basic principle in the safety system dealt with in this paper (see Fig. 1) which was designed for a fast breeder reactor for supervising the fuel element temperature [Jüngst 1976, Gmeiner 1980].

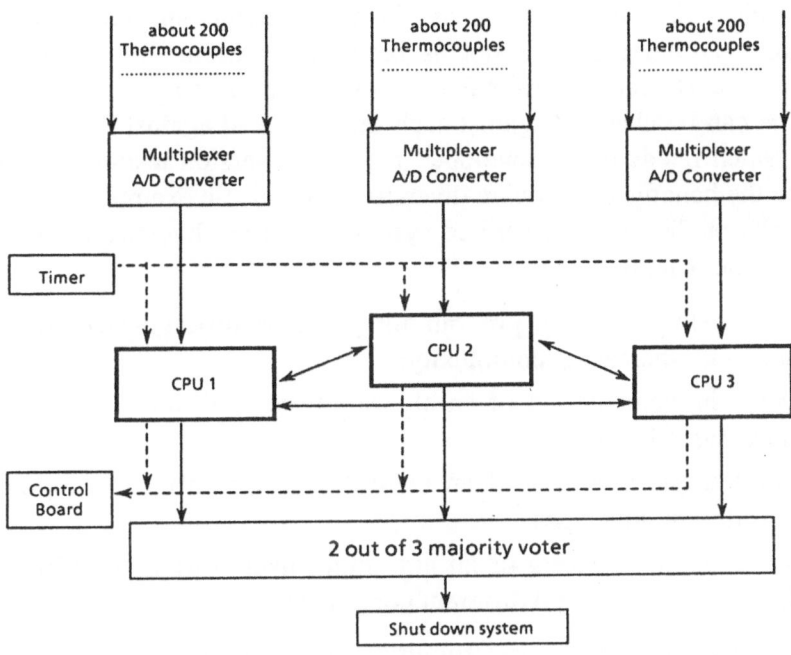

Fig. 1. Scheme of the Computerized Reactor Protection System

The reactor is equipped with about 200 thermocouples for measuring the coolant temperature, and these are triplicated for redundancy. Therefore the computer system is also a TMR-system, each line reading one redundancy line of the thermocouples. After processing the measurements in several algorithms, a data exchange takes place between the three, and then each one is performing a voting on these data and generating the final result. This in turn is voted on by a hardware 2-out-of-3 voter which is connected to the shut down system, including the control rods. The programs are executed continuously with a cycle time of about one second, triggered and synchronized by an external clock.

The original design of the system included identical software in the three computers due to cost reasons. These costs basically divide up into software development costs and validation and verification costs. With increasing availability and reliability demands especially the validation and verification costs increase. But the use of diverse software can limit this increase. On first sight one might have the impression this is wrong, especially if one looks at the software development costs, which are multiplied by the number of diverse programs. But since not all parts in the development process are done n-fold - in most cases at least the requirements specification is done only once -, total costs are not as high. In addition, the validation and verification costs can be lower than n-fold since the outputs of the diverse programs can be checked against each other instead against the output of a model which needs to be developed for single version testing. In order to evaluate the benefits of software diversity in a realistic example, we took the main parts of the above described system and used diversity in an experimental implementation.

Different levels of diverse programming can be distinguished, which are listed here with increasing complexity:

- diverse implementation of an algorithm, using different programmers, but the same language;
- diverse implementation of an algorithm, using different programmers and different languages;
- diverse implementation of an algorithm, using different programmers, different languages and different computers.

The first level of diverse programming was the basis for an investigation reported in [Avižienis 1977]. The second level will be described here. The third level of diversity does not only concern software, but also hardware. It requires an additional scale of effort. In general it can be hypothesized that the probability of common mode errors in the resulting system is decreasing if the level of diversity in the above list is increasing.

3. Description of the Experiment

3.1 Implementation

The starting point for each single implementation was a common specification. In order to have the specification as precise, consistent and complete as possible, we have chosen a formalized specification method which is similar to the input-process-output approach [Boehm 1974]. The total program is described as a set of mutual linked processes. Each of these processes consists of input, output and states, and each process state is defined by the actual values of the basic set of variables. One aspect in the specification is quite important: In addition to the synchronization points at the end of each program cycle a set of "internal checkpoints" was specified, where some intermediate results of the diverse implementations can be compared. These checkpoints give the possibility - as will be explained later - to detect certain errors in the implementations which could not be detected by simple comparison of the final program outputs.

For the implementation we used the different languages IFTRAN - an extension to FORTRAN 66 with structured programming constructs [IFTRAN 1976] - , PASCAL, and PHI2, a macro assembly language which contains structure macros for "if-then-else", "case", "do-loop" etc. [PHI2]. These languages were chosen since they were available on our SIEMENS 330 computer, and on the other hand these languages allowed to a high degree to follow the programming guidelines [Voges 1975] which were postulated for this project. Furthermore, by the use of these languages of different levels (machine level to high level) a special kind of diversity is realized, which decreases the probability of common mode errors.

The used specification method proved to support the implementations in the language IFTRAN, PASCAL, and PHI2. It was often possible to transform the specification quite naturally into the code of the programming language. On the other hand, this reduced the possibility of diversity: the programmers did not have much freedom to choose different solutions, except as forced by the languages.

The validation method [Geiger 1979], which was the basis of this experiment, contained not only constructive methods, but also analytical methods, and here mainly an intensive program test. These tests were carried out in several phases. First the program was tested by the programmer himself. Afterwards the program was retested by another person, according to predefined testing criteria. Besides diverse program development an additional aspect of diversity was realized by this technique: a diversity of the staff during the program development cycle.

As far as possible, the tests made use of automated tools like the test systems RXVP [RXVP 1985] and SADAT [Voges 1980] and an automatic result comparitor.

An additional diversity feature was the use of teams with different educational background (see Fig. 2). The system was specified by a physicist (A). The three implementations, including the first program test, were made by a computer scientist (B), an engineer (C), and a mathematician (D). The code inspection and test was performed by shifting the programs between the three implementors, and the final system test (acceptance test) was done by the specifier again.

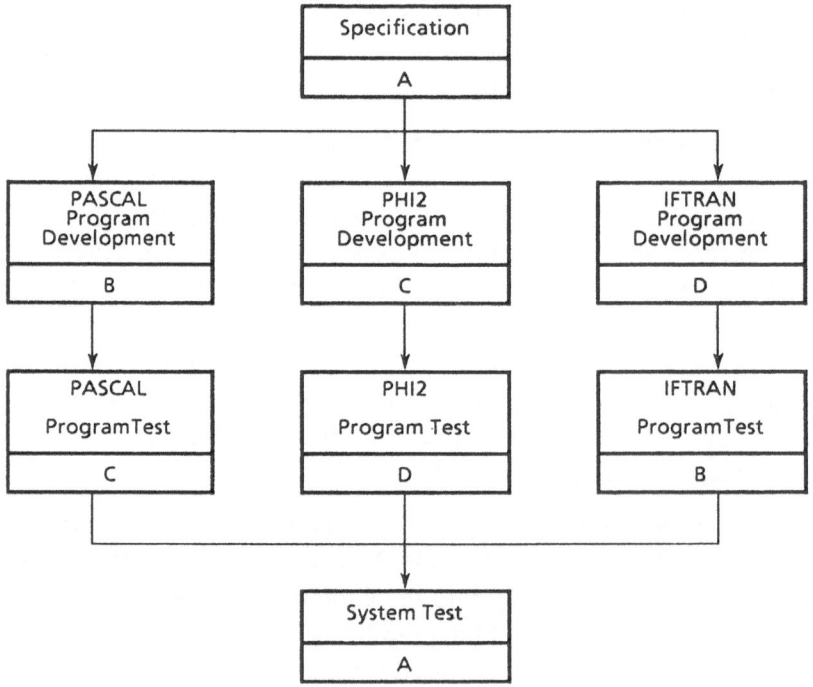

Fig. 2. System Development and Test

A summary of some characteristic implementation quantities is given in Fig. 3. The size of the specification was about 50 pages.

	IFTRAN	PASCAL	PHI2
Code size (words)	21 251	13 463	14 205
Program lines without comments	1 077	783	1 606
Run time normalized	1.21	3.65	1
Test runs before acceptance	30	42	51

Fig. 3. Characteristic Figures of the Three Implementations

3.2 Experiences

Besides the aforementioned use of tools during testing another main effort was to write down a detailed error report and analysis for the total experiment. All errors which were detected during the complete project, starting from specification, were documented on error report forms and analyzed at the end of the project [Gmeiner 1978]. In the following some of the main results of this analysis shall be explained.

Fig. 4 shows in form of a graph the relation between the error cause (arrow-tail) and its detection (arrow-head) concerning the phases of the program development cycle. 12 of the errors made in the specification were detected in the specification phase itself (review etc.), another 12 during the design, 16 during implementation and finally 10 only during acceptance testing. This demonstrates that the specification plays a dominant role. Most of the errors are closely related to the specification, and the errors induced during this phase become effective in all following phases. The main error cause was ambiguity and misinterpretation of the specification. Fig. 5, another representation of the data of Fig. 4, clearly shows that almost half of the errors are caused by the specification, but even during testing a not neglect-able amount of errors are made. Test phase errors were errors in the test frame and erroneous manual calculation of test data.

Fig. 6 gives in form of a matrix the relation between the error types and their detection methods. The total number of errors seems to be higher in Fig. 6, because some errors are counted twice if they were detected independently by different methods. Out of the used methods especially the method of "automatic result comparison" is interesting, since it is a characteristic feature for the usefulness of diverse programming. As already mentioned in

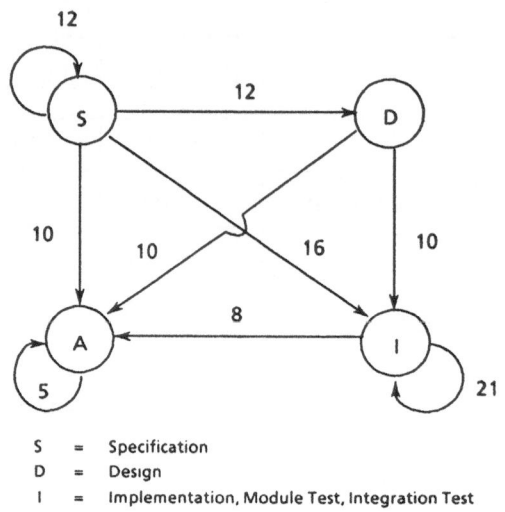

S = Specification
D = Design
I = Implementation, Module Test, Integration Test
A = Acceptance Test (by approving Authority , by User)

Fig. 4. Correlation Between Error Cause and Error Detection

Error source	Error detection in phase				Total errors in phase
	Spec.	Design	Impl.	Acc.T.	
Specification	12	12	16	10	50
Design	-	-	10	10	20
Implementation	-	-	21	8	29
Acceptance Test	-	-	-	5	5
Σ	12	12	47	33	104

Fig. 5. Error Source and Error Detection

the previous chapter, the specification defined some intermediate checkpoints, which were used during testing. An automated comparison program
checked whether the intermediate results of the three implementations at
these checkpoints were identical or not. Most of the errors which were
detected by this method were due to ambiguous specification. The following

example shall explain such an ambiguity and misinterpretation.

Error type	Detection method						
	automatic result com- parison	manual result com- parison	desk check	automatic test systems	run time system (PASCAL)	(macro) assembler, linkage editor	machine based debugging tools
incomplete specification	2	3	16	-	-	-	-
ambiguous specification	3	2	10	-	-	-	-
logical error	4	3	11	-	-	-	10
typing error	1	2	-	4	-	1	-
interface error	-	-	1	2	-	1	-
other implementation error	3	-	6	4	-	-	7
error in test frame	-	-	1	-	-	1	-
error in system software	2	4	-	-	3	2	1
other errors	3	13	-	-	2	4	1

Fig. 6. Relation Between Error Type and Error Detection Method

The specification contained the following line:

$$m^{min} \le m_i, m_j \le m^{max} \quad \forall\, i=1,...,205, \quad \forall\, j=1,...,3$$

The interpretation which was implemented by one programmer was:

$$m^{min} \le m_i \quad \forall\, i=1,...,205$$

$$m_j \le m^{max} \quad \forall\, j=1,...,3$$

while another programmer had the following interpretation, which was also the intended one:

$$m^{min} \le m_i \le m^{max} \quad \forall\, i=1,...,205$$

$$m^{min} \le m_j \le m^{max} \quad \forall\, j=1,...,3$$

These different interpretations were detected during code inspection by the person who used the second interpretation and who inspected the code with the first interpretation.

A further advantage of diverse programming is the simplification of the test process. Normally for all test input data the corresponding test output data have to be computed more or less manually. If a diverse implementation exists, the results of the two implementations can be checked against each other, and all errors resulting in different intermediate or final results can be detected. Thereby it is possible to check the diverse programs just by automatically generating large amounts of test data and by comparing the results whether they are identical or not. Only if discrepancies are detected, it is necessary to check which of the results is incorrect. But it has to be noted that it is not sufficient to use this method alone. All common mode errors, whose existence can not be neglected, are undetectable by this method. The application of other error detection methods and tests is there-fore necessary.

From the total number of detected errors 18 errors ($\cong 14\%$) were detected by the automatic comparison program. This seems to be a fairly high number, especially if one takes into account that all the other testing methods were used before the three implementations were compared with each other. Therefore these errors were probably the hard-to-detect ones.

To conclude it can be said that the automatic comparison program was a valuable tool for the comparison of the large amount of data. If this com-parison and evaluation had been done by manual inspection, much more time would have been needed, and the quality of the comparison probably would be lower with some chance of overlooking errors.

Another result showed up during the comparison testing: not only the final binary results of the three versions were compared, but also some intermedi-ate results in the form of arrays. While the binary results agreed, differences in the intermediate results were detected which led to the detection of errors in the program. This demonstrated the value of intermediate checking and the danger of checking only on a binary level with reduced information.

The use of diversity made the detection of several errors possible, and this is true not only for implementation errors, but also for specification errors. Of the more than 100 error reports collected during the experiment, no error was identical in all three implementations, and less than ten were the same in two implementations.

This experimental system was only a fragment of the complete reactor shut down system. It was not intended to go into operation. The tests were con-ducted in a simulated environment, running the three versions sequentially on one computer (Fig. 7).

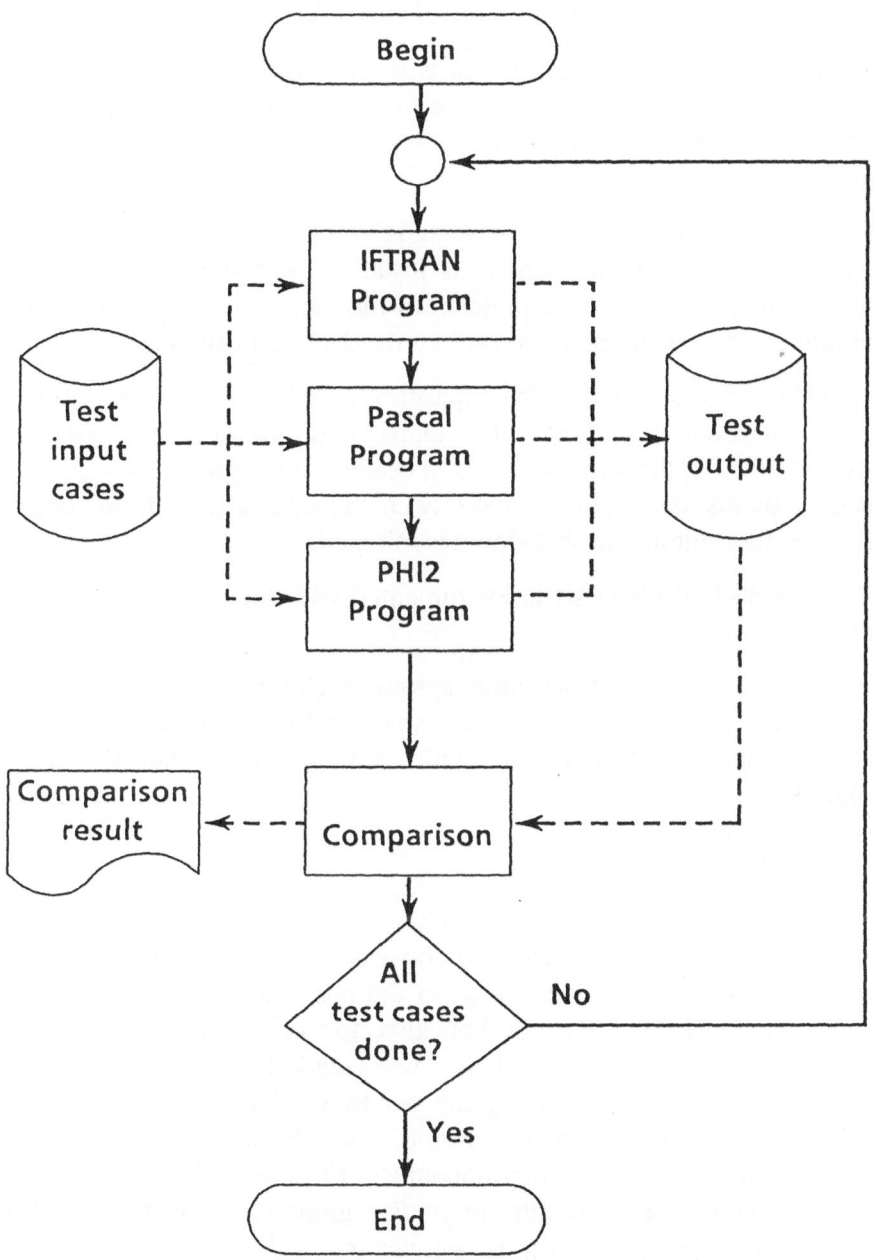

Fig. 7. Test Execution of the BPI-Programs

4. Design of MIRA

Following these positive experiences with software diversity in the above described experiment, the development of another reactor safety shut-down system was started, which included provisions for the use of diversity [Voges 1982, Voges 1985].

At the liquid metal fast breeder reactor KNK II at Karlsruhe, Germany, a hardwired core surveillance system is installed. Our aim was to substitute one part of this system, the supervision of the individual fuel element outlet temperatures of the coolant, by a computerized system called MIRA.

The purpose of the system is the supervision of the local coolant temperatures at each of the 35 individual fuel elements in order to detect local cooling disturbances. If the temperature is exceeding some predetermined or calculated set points, depending on the level a message is given to the operator or an immediate automatic shut-down is initiated.

Each of the 35 fuel elements is instrumented with three redundant thermocouples.

Since this system is part of the safety system, high reliability and availability requirements are set. This results in a high effort for hardware and especially for software design and verification which will be more explained in the following.

4.1 Hardware

The system is subdivided into four subsystems, each of which with a different function. Each subsystem itself consists of three redundant microcomputers. There are several links between the computers which are not only used for information exchange but also for error-detection and fault-tolerance. Some fault tree analysis has been made for several designs with different levels of interconnecting the net [Schriefer 1983]. The resulting structure of the system is shown in Fig. 8. The 35 core positions are instrumented with three redundant thermocouples. Each redundancy group sends its measurements to one computer in the first group (M-1, M-2, M-3). These computers make the analog/digital conversion and then interchange their inputs. After having all three redundant measurements in digital form, each M-computer checks the quality of the measurements and calculates the mean value for each position. The mean values are sent to the next two groups (A and F).

The second group (A-1, A-2, A-3) takes the actual mean temperatures, calculates the floating set points and evaluates whether at any position this floating set point or a fixed set point is exceeded. The results of this evaluation are given to the fourth group (K).

As the second group, the third group (F-1, F-2, F-3) takes the actual mean temperatures, but in addition it takes the mean local and group temperatures integrated over a certain time period, thereby taking into account the history and the trend, too. Again the floating set points are calculated, an evaluation is made and its results are given to the fourth group (K).

The fourth group (K-1, K-2, K-3) makes a final combined evaluation of the results of group A and group F. A majority vote is made for each position. If at any position an unallowed temperature is detected, automatic scram is initiated by output of a "0"; if everything is within the limits, a "1" is the output, and operation can continue.

This final signal is then used by the hardware 2-out-of-3 voter, to act on the control rods. This voter is designed inherently fail-safe and with high reliability, and it is not triplicated.

The Go=1/Stop=0 signal combination has also a fail-safe feature. In case a system is not operating, it does not produce any output and therefore its vote is taken to be "Stop". The dangerous situation arises only in the case when a system is constantly producing an output of "1".

Each of the four function groups M, A, F and K is triplicated for availability and reliability reasons. The separation of the four functions into individual units was made mainly because

-    at the beginning of the project the capacity of the available microprocessors was limited,

-    the design should be extensible for larger reactors and more limit algorithms,

-    the functions A and F should be separated because different testing strategies are needed, and

-    communication of intermediate results should be possible to allow cross-check-points with comparison of the diverse software.

The hardware of the twelve microcomputers within the system is mainly identical; the main difference is the amount of I/O-ports due to the different interconnections. For our realization we have chosen the Siemens SMP E8, a single board computer based on the Intel 8088 microprocessor chip. This microcomputer was selected due to the good price-performance relation and the small and compact board design. In addition the indirect connections to

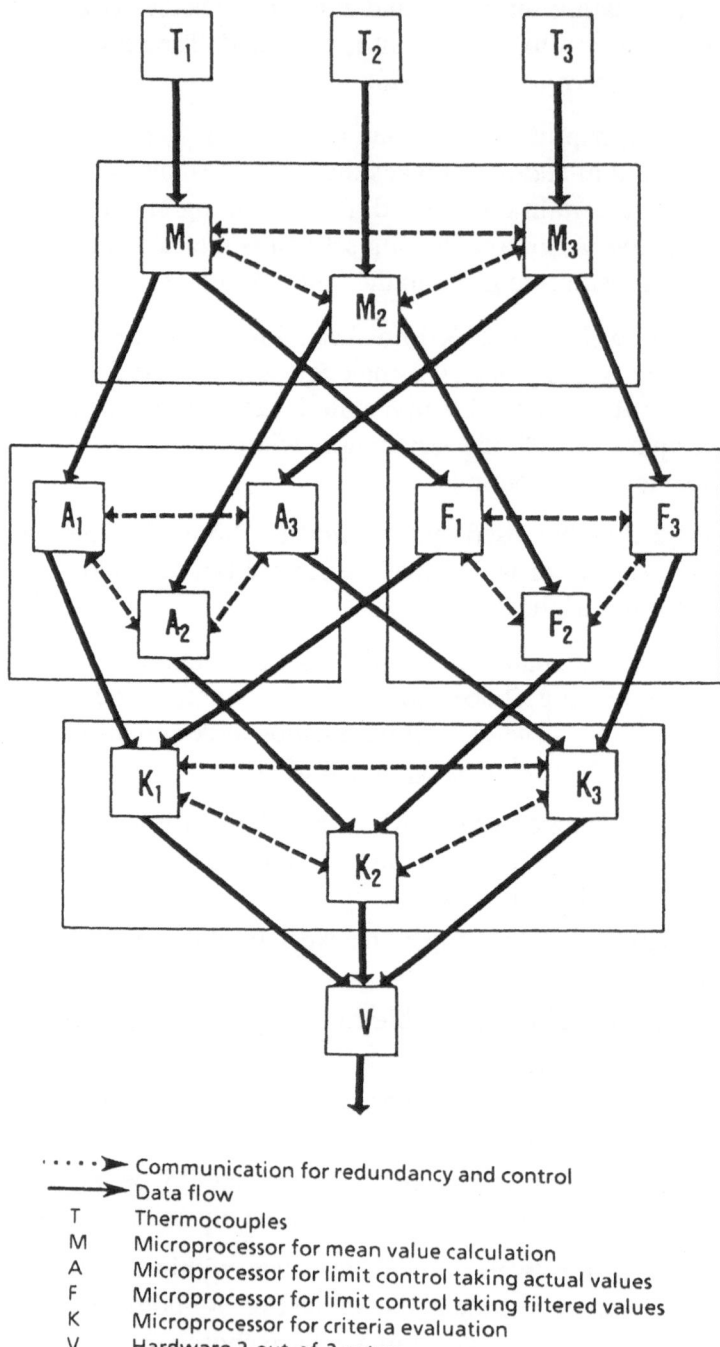

Fig. 8. Structure of the Computerized Protection System MIRA

the bus via plugs are features which are valuable in the licensing procedure.

Each single board computer consists of
- chassis with power supply
- CPU
- 64k byte main memory (EPROM and RAM)
- I/O interfaces (serial and parallel).

We use no external storage device or other peripheral units in this kernel system. Only for maintenance there is a possibility to connect some diagnostic aids to an otherwise unused I/O-port.

The interconnections between the microcomputers are not done via a common bus, but by single point to point connections, to reduce the error propagation possibility. The connections within a redundancy (e.g. M-1⇒A-1, M-1⇒F-1, F-1⇒K-1,...) are parallel connections, while the connections between the different redundancies (e.g. M-1⇔M-2, F-2⇔F-3,...) are using fiber optic links in order to provide independence and to avoid electrical interference.

In order to provide the operators with some additional information, a separate system is set up which gets basic information from this system, and which keeps records of the data and displays the actual information in different forms on color VDU screens [Elies 1984].

4.2 Software

The main parts of the software for the microcomputers are
- supervisor/operating system,
- communication software,
- application software, and
- self-testing software.

The operating system contains only the parts needed for this application. It is no special development, but a standard operating system from the manufacturer of the hardware. For the licensing, credit will be taken from other commercial use of this operating system. Main parts in this system will be the I/O-drivers and the task-management. Since several tasks are running in one processor, the correct scheduling and also the time are important subjects.

The communication software incorporates the protocol mechanism and also features for error detection in the transmission. The protocol itself is rather simple, since we have only point to point connections and also fixed length blocks on most lines. The raw data are extended with some redundancy for error detection as well as time stamps.

The application software consists of the actual problem related programs. This is the part which will vary in the twelve microcomputers, while the other software will be identical in all systems. The basic functional description of the four sets was given in the previous section.

The self-testing software runs in the spare time of each computer. Its purpose is to control the correct functioning of the hardware, to check the constant part of the memory, and to test the variable part of the memory. The results of these checks are reported to the central protocol unit as well as to the next computer in line. By these self-checks errors shall be detected as soon as possible in order to have a very low probability of double error within one unit between maintenance phases.

4.3 Fault Avoidance

During the software development process, everything has to be done in order to produce dependable software. Certain guidelines have to be followed, and there has to be some evaluation, to find out whether the final product achieves the goals set or not. In the following, we will explain the related constructive and analytical methods.

Constructive methods. A main problem in the software development is the specification. Past experience as in the above described experiment shows that many errors are introduced in this phase, and that these are the errors which are costly to remove. Some relief is seen if formal specification techniques are applied. Therefore the software requirements specification and the design use a formal specification technique which is supported by a tool. This tool checks the consistency and the completeness of the specification as well as its syntactical correctness [Eckert 1981].

The programming itself will follow programming guidelines which reflect the state of the art in software engineering [EWICS 1981]. The programming languages themselves support the ideas of structured programming.

Analytical methods. We can not only trust the success of the constructive methods, but we have to prove the correctness of the software. Therefore different analytical methods have to be applied parallel to the development of the code as well as after completion of the coding. This includes the implementation of a software quality assurance plan [IEEE 1984].

Main parts of this plan are reviews and tests, which will also involve the licensing authorities. The reviews have to start at the very beginning with the first document produced, the requirements specification. A check is necessary, whether it complies with the global system requirements, in addition to the tests done with the specification tool.

In later phases of the project, the new documents are compared with the results of the previous phase which were the input to this phase. Consistency, completeness and correctness are the main points which are looked at.

The final testing of the code is done in several steps, starting with module test, then program test, and system test. Each test involves different test data sources as well as different testing personnel. Statistics are gathered to control the success of the testing effort [Voges 1983].

4.4 Fault Tolerance

In addition to the above mentioned fault avoidance techniques something more has to be applied in order to cope with errors. For hardware redundancy, e.g. replication of modules is used, and diversity is the approach we use for software.

Several means to achieve diverse software were planned, e.g.:

- different development teams,
- different programming languages, and
- different tool sources.

In addition during testing different test data sources should be used for achieving a broader test coverage.

Within our project the programs should be written in IFTRAN/FORTRAN, PASCAL, and PL/M, by different teams with no direct contact to each other. The functional description of the program, that is the software requirements specification, is identical in all three cases. No additional algorithm diversity as a means of forced diversity is anticipated.

Experience from our previous experiment mentioned above as well as from other projects [Avižienis 1984, Bishop 1987] shows that the amount of identical errors in diverse programs is much smaller than the overall amount of errors in a single program. If a high amount of testing is done, the largest portion of the errors will be detected, even of the common source errors. In addition, use of diverse software during normal operation and not only during testing will provide an on-line error-detection possibility.

The positive effects of diverse software are not limited to the tolerance of software errors itself, but can be extended to the underlying system software and hardware: since the software is so different, the probability of concurrent activation of a common error in the system software or hardware is lower than in the case where identical software was used. This demonstrates that software diversity increases the overall system reliability.

Additional on-line error-detection techniques applied in this system include plausibility checks, control data, information exchange and error checking, and the self-checking already mentioned earlier.

All incoming data are checked for plausibility that is whether they are within the expected range. E.g., the outlet temperatures may not be lower than the inlet temperature. If this is the case, it is due to an instrumentation failure or a transmission error.

Between the real measurement data, some additional control data are interspersed. These are processed in the same manner as the normal data and are handled as if they were additional reactor core positions. The final evaluating group (K-computers) checks, whether the results at these positions are the predetermined, expected ones or not. If they deviate, they reveal some irregularities in the algorithm or other processing parts, which could have effected the normal data, too. Therefore a message is given to the operators, and some fail-safe action is necessary.

Within each functional group an information exchange between the redundancies is conducted. The two main reasons for this are error-detection and fault-tolerance. E.g., F-1 receives the output of M-1 directly from M-1, the output of M-2 indirectly via F-2, and the output of M-3 via F-3. These outputs are compared. All detected discrepancies are reported to the protocol units. These can evaluate the messages and detect the source of the discrepancies, whether it is an M-unit, or a transmission line between the F-units, e.g. The status of the reactor with respect to the temperatures and of the computer system itself is displayed and maintenance is alert in case of failures.

If the directly received data are considered to be erroneous, one of the indirectly received but correct data blocks can be used for further processing. Therefore single failures on each level can easily be tolerated without remarkably degrading the complete system. In addition, this comparison of intermediate results allows error detection between the diverse programs which could be invisible at the final voting comparison, which is on only one bit.

4.5 Status

The hardware of the system was installed off-site and is undergoing testing since mid 1983. It is connected with the reactor core instrumentation via interfaces and a 1.5 km fiber optic cable. It receives the original reactor measurements on-line, but is not activating the control rods. The tests resulted in some redesign of the system, which were assisted by the availability of new hardware components like fiber optic links. The testing showed that the overall reliability of the hardware was higher than the reliability data provided by the manufacturer and based on MIL STD 217C. The data were collected over a period of three years with about fifteen computer running (including the testing interface). Only a prototype of the software is in the test now. The further development, including the diverse programs and the licensing procedure, was cancelled due to new priorities within our research institute. Therefore, the old hardwired shut-down system will continue to run.

5. Conclusion

The probability of a common mode failure (the same error identical in all implementations) is very low if the teams work independently, the languages have different levels (assembly language and problem oriented language, e.g.) and the compilers were not designed by the same people. The most probable reason for a common mode failure is an error in the specification. Therefore much effort has to be put into correct, consistent and complete formal specification. However, there is a problem in detailing the specifications too much: detailed specification leads to similar internal program structures of the different implementations as was seen in our first experiment. Thus the advantage of diverse programming gets partly lost.

According to our experience one of the main problems in diverse programming is the synchronization of the different programs, especially if the programs have intermediate checkpoints. These checkpoints have to be designed explicitly, very carefully and in a detailed manner. They proved to be valuable especially since we detected a few errors, where final results agreed and only the intermediate results disagreed. In these cases we have detected some hidden errors, not activated directly by the test cases.

Furthermore, the experiment showed that diverse programming detects certain errors which are unlikely to be detected with other methods.

References

[Avižienis 1977]   A. Avižienis and L. Chen: On the Implementation of N-Version Programming for Software Fault-Tolerance during Program Execution. Proceedings COMP-SAC '77, 1977, pp. 149-155.

[Avižienis 1984]   A. Avižienis and J.P.J. Kelly: Fault Tolerance by Design Diversity: Concepts and Experiments. IEEE Computer Vol. 17 (August 1984) 8, pp. 67-80.

[Bishop 1987]   P.G. Bishop: The PODS Diversity Experiment, this book.

[Boehm 1974]   B.W. Boehm: Some Steps toward Formal and Automated Aids to Software Requirements Analysis and Design. 2nd IFIP Congress, Stockholm, 1974.

[Eckert 1981]   K. Eckert and J. Ludewig: ESPRESO-W - Ein Werkzeug für die Spezifikation von Prozeßrechner-Software. In: G. Goos (Ed.) Werkzeuge der Programmiertechnik, Berlin, Springer-Verlag Berlin-Heidelberg-New York 1981, pp. 101-112.

[Elies 1984]   V. Elies: A Protocol System as an Extension of the MIRA Reactor Protection System. IAEA-Meeting Saclay, F, 1984.

[EWICS 1981]   EWICS: Development of Safety Related Software. EWICS TC7 Position Paper No. 268, 1981.

[Geiger 1979]   W. Geiger, L. Gmeiner, H. Trauboth and U. Voges: Program Testing Techniques For Nuclear Reactor Protection Systems. IEEE Computer 12 (August 1979) 8, pp. 10-18.

[Gmeiner 1978]   L. Gmeiner: Projektbegleitende Fehleraufzeichnung und -auswertung während der BESSY-Pilotimplementierung (unpublished 1978).

[Gmeiner 1980]   L. Gmeiner and U. Voges: Software Diversity in Reactor Protection Systems: An Experiment. Proc. IFAC Workshop SAFECOMP'79, Oxford, Pergamon Press 1980, pp. 75-79.

[IEEE 1984]   IEEE: Standard for Software Quality Assurance Plans, IEEE Std 730, 1984.

[IFTRAN 1976]   Structured Programming Preprocessors for FORTRAN. General Research Corporation, Santa Barbara, 1976.

[Jüngst 1976] U. Jüngst: Design Features of the Fuel Element Computerized Protection System. IAEA/NPPCI Specialists' Meeting, München, 1976.

[PHI2] PHI2-Programmierhilfe-Makros für strukturierte Programmierung. SIEMENS Programmbeschreibung P71100-J1015-X-X-35.

[RXVP 1985]   RXVP 80. The Verification and Validation System for FORTRAN. User's Manual. General Research Corporation, Santa Barbara, 1985.

[Schriefer 1983] D. Schriefer, U. Voges and G. Weber: Design and Construction of a Reliable Microcomputer-Based LMFBR Protection System. In: Proceedings of Internat. Workshop on Nuclear Power Plant Control and Instrumentation, IAEA-SM 265, 1983, pp. 355-366.

[Voges 1975]   U. Voges and W. Ehrenberger: Vorschläge zu Programmierrichtlinien für ein Reaktorschutzsystem. KfK-Ext. 13/75-2, Kernforschungszentrum Karlsruhe, 1975.

[Voges 1980]   U. Voges, L. Gmeiner and A. von Mayrhauser: SADAT - An Automated Testing Tool. IEEE Trans. Softw. Eng. SE-6 (May 1980) 3, pp. 286-290.

[Voges 1982]   U. Voges, F. Fetsch and L. Gmeiner: Use of Microprocessors in a Safety-Oriented Reactor Shut-Down System. EUROCON '82, Lyngby, DK, 14-18 June 1982. E. Lauger, J. Moeltoft (Eds.), Reliability in Electrical and Electronic Components and Systems. Amsterdam: North Holland Publ. Co. 1982, pp. 493-497.

[Voges 1983]   U. Voges and J. R. Taylor: Systematic Software Testing, In: Proceedings of EWICS, Schriftenreihe der Österreichischen Computer-Gesellschaft 21 (1983), pp. 165-183.

[Voges 1985]   U. Voges: Application of a Fault-Tolerant Microprocessor-Based Core Surveillance System in a German Fast Breeder Reactor. EPRI-Seminar: Power Plant Digital Control and Fault-Tolerant Microcomputers, Scottsdale, AZ, USA, 9-12 April 1985.

The PODS Diversity Experiment

P.G. Bishop
Central Electricity Generating Board
Central Electricity Research Laboratories
Leatherhead
Surrey KT22 7SE
England

1. Introduction

A high integrity system typically has a number of redundant components operating in parallel to reduce the probability of a system failure. If the component failures were random, then the probability of several components failing simultaneously would be much smaller than the failure probability of any single component. However, should the components contain common design flaws, then more than one component could fail simultaneously due to a common cause (a common mode failure). This would increase the probability of a system failure. For a computer-based system where the same software "component" is being run in each processor, any software fault is a potential cause of common mode failure. One method of reducing common software faults is to use diverse software in each processor (n-version programming [Avižienis 1975]).

The use of software diversity raises a number of issues. Do diverse programs contain independent faults or are the faults correlated? Would the extra effort required to produce diverse software be better spent in attempting to produce one correct software component? Is the quality of the initial specification more important than a diverse implementation? The Project on

Diverse Software (PODS) was, in part, set up in order to provide some insight into these questions.

PODS [Barnes 1985, Bishop 1986] was a project between the Safety and Reliability Directorate (SRD) and the Central Electricity Research Laboratories of the Central Electricity Generating Board (CEGB) in England, the Technical Research Centre of Finland (VTT) and the Halden Reactor Project (HRP) in Norway. The purpose of the project was to determine the effect of a number of different software development techniques on software reliability. The main objectives were:

- To evaluate the merits of using diverse software.
- To evaluate the specification language X-SPEX [Dahll 1983].
- To compare the productivity and reliability associated with high-level and low-level languages.

In addition, there was a secondary objective to monitor the software development process, with particular reference to the creation and detection of software faults.

2. Experimental Design

To achieve these objectives, an experiment was mounted which simulated a normal software development process to produce three diverse programs to the same requirement. The requirement was for a reactor over-power protection (trip) system. After careful independent development and testing, the three programs were tested against each other to locate residual faults. All phases of the project were carefully documented for subsequent analysis.

The structure of the project is summarized in Fig. 1. SRD acted as the "customer" while CEGB, HRP and VTT each took the role of software "manufacturers". The three manufacturers produced their programs independently of each other. Further diversity was introduced by placing different implementation constraints on the manufacturers. Different combinations of programming language (Fortran 77 or Nord assembly language) and specification methods (the X language and informal) had to be used by each team. This also made it possible to compare the different specification techniques and programming languages. An extra element of diversity was introduced by supplying two different reactor power calculation algorithms. The constraints were applied in a specific pattern, as shown in Table I, so that the results of each factor could be assessed individually.

Two different manufacturer's specifications were independently produced

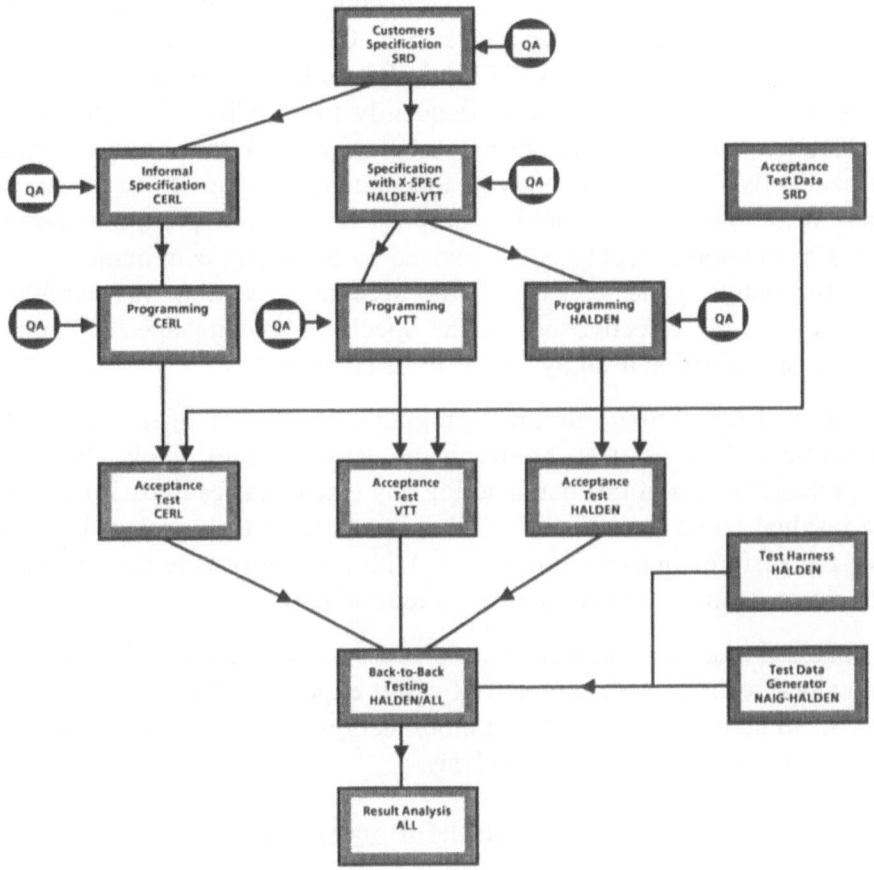

Fig. 1. PODS Project Organization

Table I. Implementation Constraints for Each Team

Team	Program Language	Main Algorithm	Spec. Method
CEGB	Fortran	Polynomial	Informal
HRP	Assembly	Table	X
VTT	Fortran	Table	X

from the customer specification, each reflecting the manufacturers' interpretations of the customer's specification. CEGB produced an informal specification, while HRP and VTT jointly produced a specification in the language "X". HRP and VTT subsequently took individual copies of this specification, and the three teams then developed their own programs independently. Each team followed the same development procedure, consisting of design and code production phases. All three programs then had to pass a common acceptance test devised by SRD. No communication was permitted between the teams until the acceptance testing was completed. This allowed the effectiveness of the specification language "X" and the informal specification used by CEGB to be compared.

After acceptance testing, the three programs were tested against each other "back-to-back" in a test environment which could apply large and comprehensive sets of test data and log any discrepancies in their responses. The residual faults discovered by this process were then analysed to determine the performance of the programs with a majority vote for comparison with the performances of the programs individually.

To enable the software development process to be observed, all participants maintained records of the amount of effort expended in each of the project phases. In addition, all faults and modifications were recorded on specially designed forms for subsequent analysis.

3. Project Management

To ensure that the maximum amount of information was gained from the experiment the whole project was comprehensively documented and clear responsibilities where defined for all the team members as shown in Table II.

SRD was responsible for the overall management of the project, and produced a formal project document describing the:
- project objectives,
- project tasks,
- project phases and time-scales
- documentation required,
- document referencing standards,
- fault reporting formats,
- fault classification scheme,

Table II. Project Members and Responsibilities

Company	Member	Project Activities
SRD	M. Barnes P. Humphreys A. Ball	Project Manager Global QA, Cust. Spec Acceptance Test Design
CEGB	P. Bishop D. Esp P. Rutter	Man. Spec, Local Manager Man. Spec, Design, Code Local QA
HRP	G. Dahll O. Hatlevoldt S. Yoshimura	X-Spec, Local Manager Test Harness Design, Code Test Strategy, Local QA
VTT	J. Lahti B. Bjarland	Local Manager, X Spec QA Local QA Design, Code

- software change control procedure,

- effort recording format.

Further documents defined the quality control procedures and the testing strategies. Some of the major aspects of the project management are discussed in the following sections.

3.1 Project Phases

The project was organised into phases to provide a structured approach to project control. The phases comprised:

- customer specification,

- manufacturer specification,

- design,

- coding,

- acceptance testing,

- back-to-back testing.

The end of each phase was regarded as a "milestone" in the progress of the project. No phase was considered to be complete until all the documentation required for the phase was released. Any subsequent faults discovered in a released item had to be documented and processed using a formal change control procedure. Due to an oversight, the project control document did not specify that faults occurring **within** the manufacturers' development phases should also be recorded. In the event, both CEGB and HRP did record this information, but VTT did not. As a result, some of the analyses of the software development process had to be restricted to the CEGB and HRP teams.

3.2 Quality Control

Quality assurance guidelines, based upon the EEA quality assurance guide [EEA 1981], were issued to each team by SRD. The guidelines provided the ground rules from which each team derived its own quality assurance procedures. During the software development, each team had to produce the following documents:

- local project control plan,
- local quality assurance plan,
- description of work done in each phase,
- quality assurance applied to each phase,
- design document,
- description of design methodology,
- man-hour and fault report documentation.

An independent quality assurance representative was nominated for each development site to verify that the software development conformed to the agreed project standards.

3.3 Test Documentation

During the acceptance and back-to-back test phases the following information was recorded:

- copies of all the different program versions,
- fault reports and change notes,
- copies of all test data sets for acceptance and back-to-back testing,
- logs of all failures,
- listings of each version of the test harness program,
- man-hours of effort

This level of documentation makes it possible to repeat existing tests, or conduct new tests on any of the post-development program versions.

3.4 Fault Classification Scheme

A structured fault reporting scheme was devised which permitted faults to be described from different viewpoints, (e.g. the cause of the fault, the effect, the detection method etc). The form was designed to be computer-based. It had a regular syntax where the main categories and sub-categories were defined by key-words. The user could also include unstructured information using text strings. An example form is shown in Fig. 2. The form was accompanied by a document defining the fault reporting syntax and the meaning of each term.

4. The Customer Specification

In the customer specification phase, SRD produced an informal customer's requirement specification (PODSPEC). This specified a realistic reactor trip system, incorporating the following functions:

- reactor power calculation based on analogue input data,
- calibration using encoded thumb-wheel switch data,
- primary and secondary trip logic,
- test mode,
- ROM integrity check,
- diagnostic indications,
- initialization on request.

Fig. 3 summarizes the main data flows and processing functions which had to be implemented.

The customer specification also contained the following performance requirements:

- a maximum response time of 500 milliseconds from a change in the primary input signal to a corresponding change in the output signal,
- a failure to trip on demand of no more than 1 in 10,000.

SRD created and maintained separate versions of their requirement specification for each team. These versions were independently modified when faults were reported by the teams to prevent any unintended transfer of information during the development process. This independence of fault

```
----------------------------------------------------------------------------
                              PODS ERROR RECORD
IDENTIFICATION
    local ref number      : PERAC 100eg
    master ref number     : ???
    title                 : DL / range error flag
    version               : 4
    company               : CERL
    reporter              : DGE
CREATION
    item type             : DESIGN
    item name             : CERLDES
    item version          : 1
    item internal ref     : N-A
    phase                 : DESIGN          /date: 17-10-83        /time: ???
    mental activity       : ANALYSE
    operation on item     : CREATE
    error classification  : DOCUMENT:CONTENT:INCORRECT
    error description `    : "In the design, the analog input/output range error
                            flag (RE) was not implemented according to its
                            CERLSPEC definition.  The effect of a range error
                            in output DL is not taken into account.
                            CERLDES version 0 did include DL range error
                            in RE, and it was intended to perpetuate it in
                            CERLDES version 1 (see PERDC 120) but it was
                            omitted from CERLDES 1 in the event.
    error cause           : HUMAN:FORGET
DISCOVERY
    item type             : PROGRAM
    item name             : CERLPROG
    item version          : 1.2.4
    item internal ref     : N-A
    phase                 : LACTST          /date: 15-11-83   /time: ???
    discoverer            : DGE
    method                : TEST:ACCEPT:ACCTEST2
    error indication      : Test failures
CONSEQUENCE
    internal function     : ERROR FLAG LOGIC
    external effect       : "The analog input/output range error flag RE is not
                            set when all analogue inputs are in
                            range but DL is range-limited inside the program."
    failure type          : DANGER
CHANGE
    phase                 : LACTST          /date: 15-11-83   /time: ???
    changer               : DGE
    items changed         : DISCOVERY-ITEM
    description           : "Make range check on DL in subroutine SIGPROC
                            (where DL is calculated) and set a new flag
                            REDL accordingly.  Rename previous RE flag as
                            REIN (RE for INputs).  Make RE=REIN 'OR' REDL."
VALIDATION
    phase                 : LACTST          /date: 15-11-83   /time: ???
    validator             : DGE
    method                : TEST:ACCEPT:UNOFFICIAL:UNRECORDED
RELATED ITEMS
    detection report      : ???
    error report          :-  local ref: AC 10      /master ref: ???
    change report         :-  local ref: AC 10      /master ref: ???
    other                 : ???
----------------------------------------------------------------------------
```

Fig. 2. Example Fault Record

reporting was retained until the completion of the acceptance testing phase.

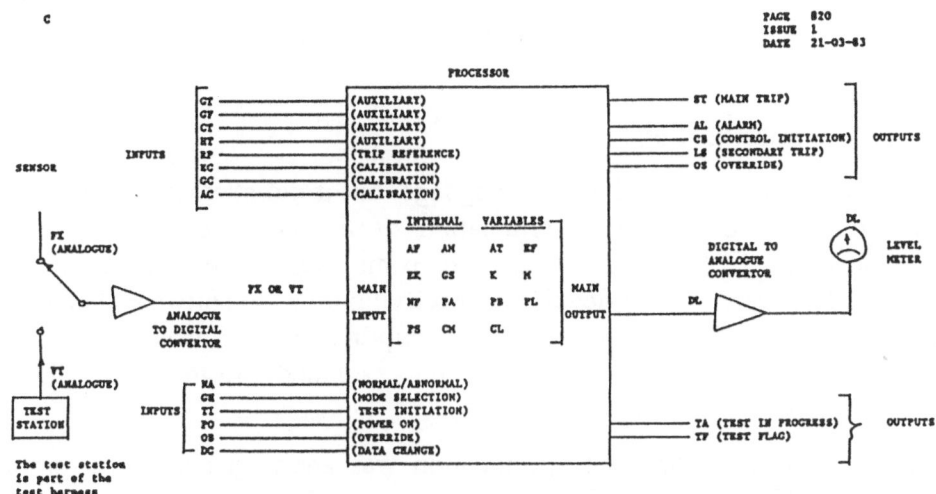

Fig. 3. Data Flows within the Trip System

5. The Manufacturers Specification

5.1 X Specification

X-SPEX [Dahll 1983] is a system which consists of two parts, the specification language X and a computerized specification tool SPEX. The aim of this system is to describe the customer requirements using a formal language (X) aided by the SPEX tool which can check for completeness and consistency in the specification. X is an entity - relationship - attribute type specification language based on RSL [Alford 1977] and adopts part of its terminology. A specification written in X consists of a set of entities with specified attributes which are linked together by relationships. Using the SPEX tool, entities, relationships and attributes can be added, modified and deleted within a database. The database can be subjected to a number of analyses to check for completeness and consistency.

The X specification was produced by members of the HRP team, while VTT performed the quality control checks. The process of writing the specification in X revealed deficiencies in the definition of the interfaces and the exact response of the the trip system to different stimuli. A number of different quality checks were applied to check for internal consistency and compatibility with the customer specification. At that stage, SPEX had not been fully developed, so some of the checks were performed manually.

After the formal release of the X specification, separate copies were issued to HRP and VTT. Subsequent revisions to the specifications were made independently by the two teams.

5.2 CEGB Specification

CEGB did not use a specification language, but it did attempt to produce a concise and unambiguous interpretation of the customer's requirement specification. CEGB considered that some aspects of the customer specification were too low-level (i.e. specifying how the software should be implemented rather than the real requirements). Most of these implementation-specific features were excluded from the CEGB specification.

The customer's requirement specification was analysed using peer-group inspections [Fagan 1976]. The trip logic was re-formulated using Boolean logic definitions and state machine diagrams [Minsky 1967] which were considered to provide a clearer and more abstract definition of the customer's requirements. This approach encouraged a higher level of analysis leading to the discovery of potential problems in the customer specification of the trip logic.

6. Software Production

All teams subsequently produced their programs based on their own specifications. Software development methods recommended by Myers [Myers 1976] were used by all teams. This involved documentation and inspection at each stage of development. All teams used a top-down design approach where a complete software design was produced and inspected before detailed design and coding commenced. The various techniques employed by the teams are summarized in Table III.

Since the X language is intended to be used at any level of detail, some aspects of the manufacturers functional specification were expanded to form a design specification. The design was then transcribed into Nassi-Shneiderman diagrams. A section of the X specification produced by HRP is shown in Fig. 4.

Table III. Summary of Software Development Methods

Activity	CEGB	HRP	VTT
Design	Yourdon-Constantine Pseudo-code	Nassi-Shneiderman diagrams X language	Flow-charts Data Flow matrix Pseudo-code
Coding	FORTRAN 77	Nord Assembly Language	FORTRAN 77
Test	Module test System tests Symbolic debug	System tests	Module test System tests Symbolic debug
QA	Inspections	Inspections	Inspections

7. Acceptance Testing

In the acceptance testing phase, each of the three programs was submitted to a common set of tests defined by the customer, SRD. The objective of the acceptance testing was to provide coverage of the input domain and the internal states of the software in an economical manner. A total of 672 individual test cases were produced, grouped into sequences of systematic tests and randomly selected test data.

7.1 Systematic Tests

Data sets were derived to test the software for compliance with the customer specification in the following areas:

- reactor power algorithm
- trip logic and alarm indications
- internal integrity checks
- calibration changes
- test mode operation

In each test set, data was generated to test for correct operation with both valid and invalid input data.

```
ALPHA: TRANSFER_INPUT_DATA.

    INPUTS:
          DATA: FXIN
          DATA: VTIN
          DATA: CTIN
          DATA: HTIN
          DATA: RPIN
          DATA: DIGIN1
          DATA: DIGIN4.

    OUTPUTS:
          DATA: FX
          DATA: VT
          DATA: CT
          DATA: HT
          DATA: RP
          DATA: GT
          DATA: GF
          DATA: NA
          DATA: TI
          DATA: OB
          DATA: INPUT_ERROR
          DATA: DIGIN4.

    REFERRED_BY:
          SUBNET: TRIP_COMPUTATION.

    DESCRIPTION:
          "THE INTERNAL DATA ARE CALCULATED FROM THE
          DATA IN THE INPUT INTERFACES ACCORDING TO THE FORMULA GIVEN
          IN THE FUNCTION BELOW. IF ANY OF.THE VARIABLES IS OUTSIDE ITS
          RANGE, AS DEFINED IN THE DATA ELEMENTS, THEN INPUT_ERROR IS
          TRUE, AND THE PARAMETER SHALL BE SET TO THE EXCEEDED RANGE LIMIT"

    ENTERED: 83/08/08.

    ENTERED_BY: G_DAHLL.

    COMPLETENESS: CHANGEABLE.

    FUNCTION:
          FX:=FXIN*160/4095-5
          VT:=VTIN*160/4095-5
          CT:=CTIN*520/4095-10
          HT:=HTIN*520/4095-10
          RP:=THE_NEAREST_INTEGER_TO(RPIN*100/4095)
          GT:="THE NUMBER FORMED BY DIGIN1[8..15]"*110/255-5
          GF:="THE NUMBER FORMED BY DIGIN1[0..5]"*110/255-5
          NA:=DIGIN4[5]
          TI:=DIGIN4[3]
          OB:=DIGIN4[1]
          DIGIN4[1]:=0
          IF (THERE_IS X, X MEMBER_OF (FX,VT,CT,HT,RP,GT,GF):
          NOT X VALUE_IN "ITS RANGE AS DEFINED IN ITS SPECIFICATION")
            X:="THE EXCEEDED RANGE LIMIT"
            INPUT_ERROR:=TRUE
          OTHERWISE
            INPUT_ERROR:=FALSE
          END
          IF (NOT RP MEMBER_OF (16,20,27,32,33,44,53,73))
           RP:="THE MEMBER_OF (16,20,27,32,33,44,53,73) WHICH IS NEAREST RP"
           INPUT_ERROR:=TRUE
          END
    END.
```

Fig. 4. Example X Specification

7.2 Random Tests

The random tests were designed to permit the variation of as many inputs as possible over their full range. A uniform distribution was used for each input parameter, giving equal weighting to the selection of any value over the full range of each parameter.

7.3 Testing

The HRP program was tested in a single stage using the NORD computer at Halden, while the CEGB and VTT acceptance testing was performed in two stages. First, both teams constructed their own local test harness which was used to apply the SRD acceptance test data. On satisfactory completion of the acceptance tests, both companies had to transfer their programs to run on the HRP NORD computer. In practice there were minor faults in the test data that were only resolved with SRD when the CEGB and VTT programs were re-accepted at Halden.

8. Back-to-Back Test Phase

Once the three programs had passed the customer acceptance tests on the HRP computer, they were tested against each other "back-to-back" to detect residual faults. The back-to-back test harness, shown schematically in Fig. 5, was designed to apply common input data to all three trip programs and then check the resultant output values. Program faults were revealed by discrepancies between the outputs of the three programs. Since the main purpose of the experiment was to detect software discrepancies between the programs the test harness was made extremely simple. Each program was implemented as a subroutine which was linked into the main test harness. Identical test data could therefore be applied to all three subroutines and the results could be compared once the three subroutines had completed. This approach avoided all the synchronisation problems that can occur if the test harness and the three trip programs were implemented as independent programs.

8.1 Strategy for Test Data Selection

The test philosophy for back-to-back testing was developed by an independent member of the HRP team. The tests were designed to be as comprehensive as possible, and comprised a series of systematic checks designed to uncover particular types of faults, and checks employing pseudo-random input data. The systematic test data values were derived using a variety of

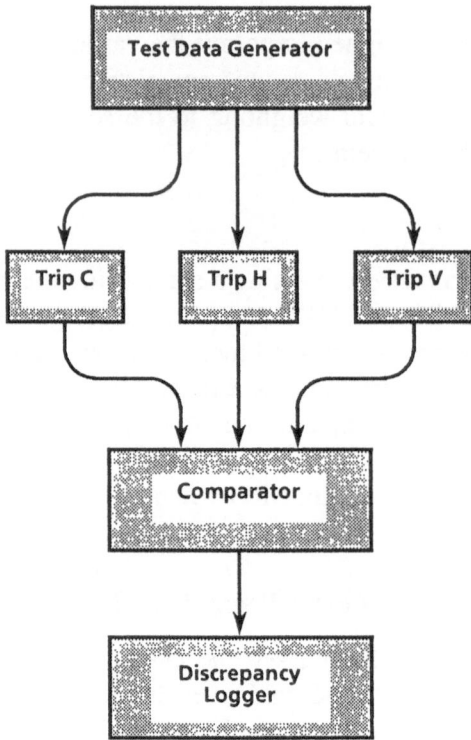

Fig. 5. The Back-to-Back Test Harness

different techniques:

- Equivalence partitioning was used to divide the input domain into a finite number of equivalence classes. Test data was designed to cover as many valid equivalence classes as possible, to maximize the effectiveness of a finite number of tests. Invalid equivalence classes (which relate to faulty input data) were designed to be tested one at a time to avoid masking effects.
- Boundary value analysis was used to derive test cases at the boundaries of input and output equivalence classes.
- Decision tables were used to take into account possible combinations of input conditions, output conditions and internal program states.

On the basis of these analyses, the following series of 2472 systematic tests were designed:

- Domain coverage check. 979 tests were constructed to test as many features of the main trip algorithm as possible.

- Hysteresis check. To avoid any "jittering" of limit check signals when the computed power is near a limit, a hysteresis band was specified. Once a limit had been triggered, it would not be cleared until the power fell below a lower limit value. 200 tests were devised to test this feature.

- ROM corruption check. 112 test cases were devised to check that the software had detected a change in the "pseudo ROM" region.

- Thumb-wheel data check. The calibration parameters were entered on thumb-wheels. The values on individual thumb-wheels were converted to internal parameters. 322 different tests were defined to check that the values were correctly converted, and that faulty input values were recognised.

- Reactor power coverage check. 859 tests cases were used to check the computation of the reactor power.

One of the advantages of software diversity is that the programs can be tested against each other using large amounts of arbitrary test data. The test harness was designed to produce arbitrary, but repeatable, sequences of test data using its own pseudo-random number generator. Test data sequences could be generated which conformed to any of the following distributions:

- Uniform over the input domain
- Gaussian over the input domain
- Rectangular around domain boundaries
- Gaussian around domain boundaries

These distributions could be applied with either a single initialization demand at the beginning of the test sequence, or with random initialization demands throughout the test.

The total number of systematic and random tests applied during back-to-back testing comprised 665,288 test cases.

8.2 Testing Procedure

A formal test procedure was adopted to ensure the correctness and consistency of testing. Whenever the programs disagreed, a group diagnosis of the cause was made, with SRD making the final judgement. The program failure together with its cause was recorded on a computer file, and the programmers completed change notices for the program modifications. The modified programs were tested against the SRD acceptance test data and, if successful, the three programs were tested together, starting again from the beginning of the back-to-back test sequence.

9. Subsequent Testing

9.1 Failure Rate Measurement Tests

Additional back-to-back test runs were performed at a later date in order to establish a relative failure rate for each individual fault detected during back-to-back testing. In these test runs, uniformly distributed random input data was applied to successive versions of each program running against a "golden" program (a final version of the trip program). It should be noted that a uniform distribution was not representative of normal operating conditions but, in principle, it was capable of activating every known fault. The failure rates for each fault were calculated from incremental changes in overall failure rate, after considering known fault interactions.

To assess the sensitivity of the failure rate to the input test data, similar tests were applied to the first post-acceptance version of each program using a pseudo-random Gaussian distribution and systematic test data.

9.2 Fault Seeding

To assess the effectiveness of the test data employed, faults were deliberately seeded into the HRP program, and then tested back-to-back against the "golden" versions of the CEGB and VTT programs using acceptance and back-to-back test data.

10. Analysis of Software Diversity

10.1 Faults Detected in Back-to-Back Testing

The faults discovered during back-to-back testing are assumed to correspond to the faults that would have been discovered during in-service operation of the trip system, since they were not detected by the acceptance tests. The assessment of the benefit of diversity is based on an analysis of these faults.

The distribution of the faults between the three teams is shown in Fig. 6. Seven different residual faults were found; six were caused by problems in the original customer specification while one related to an ambiguity in the X specification. In addition, two faults were accidentally introduced while making corrections. No faults were traced back to the implementation

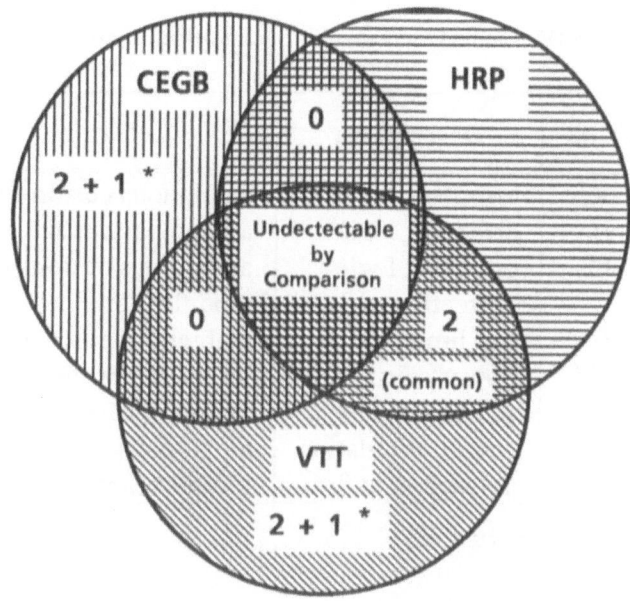

* Correction Induced Faults

Fig. 6. Distribution of Faults Detected in Back-to-Back Testing

phases.

Details of the faults discovered in back-to-back testing are shown in Table IV. Faults 6 and 7 were common to two programs and would have won a majority vote causing an overall system failure. Faults 1 and 2 were mutually exclusive and the other faults were independent, so they could only cause a failure on a single channel and would have been out-voted. The program faults marked with an asterisk have been classified as dangerous. A dangerous failure was defined as a trip or alarm signal failing to operate or the computed power level being too low.

10.2 Failure Rates of the Faults

The failure rates derived from the additional back-to-back tests using uniform pseudo-random data are shown in Table V.

This shows the probability of failure per test-case, and gives an indication of the fraction of the total input domain that is capable of activating the fault.

Table IV. Faults Discovered in Back-to-Back Testing

Fault	Team	Description	Cause
1	V	error flag logic	ambiguous customer spec
* 2	H	error flag logic	ambiguous customer spec
3	C	error flag logic	ambiguous customer spec
4	C	correction to 3	modification
* 5	C	trip limit calc	ambiguous customer spec
* 6	V,H	power calculation	incorrect customer spec
7	V,H	sec trip logic	incorrect customer spec
8	V	power calculation	ambiguous X spec
9	V	correction to 8	modification

* Fail-danger faults

Table V. Failure Rates Using Uniform Random Data

Program Versions	Failure Rate			
	CEGB	HRP	VTT	Maj. Vote
C1,H1,V1	.01	.25	.0018	.0008
C1,H1,V2	.01	.25	.0008	.0008
C1,H2,V2	.01	.0008	.0008	.0008
C2,H2,V2	.70	.0008	.0008	.0008
C3,H2,V2	.0006	.0008	.0008	.0008
C4,H2,V2	.0000015	.0008	.0008	.0008
C4,H3,V3	.0000015	.00009	.0001	.00009
C4,H4,V4	.0000015	.0000015	.0000066	.0000015
C4,H4,V5	.0000015	.0000015	.106	.0000015
C4,H4,V6	.0000015	.0000015	.0000015	.0000015

The failure rate for version 4 of the VTT program is an estimate based on a single failure observed during the main part of back-to-back testing. Unlike the other faults, it was not activated by the uniform pseudo-random test data as the number of test cycles employed was too small to reveal a fault with such a low probability of failure. The final failure rate figure of 0.0000015 is a limiting value calculated from the reciprocal of the number of back-to-back test cases employed.

The two common faults (6 and 7) had a combined failure rate of 0.0008. These faults could not be excluded by majority voting. The remaining faults were all diverse so that, although the total failure rate of all the faults in version (C1, H1, V1) was around 0.26, the failure rate after majority voting was only 0.0008. The majority vote failure rate stayed at this "plateau" level until faults 6 and 7 were removed. This effect is shown clearly in Fig. 7.

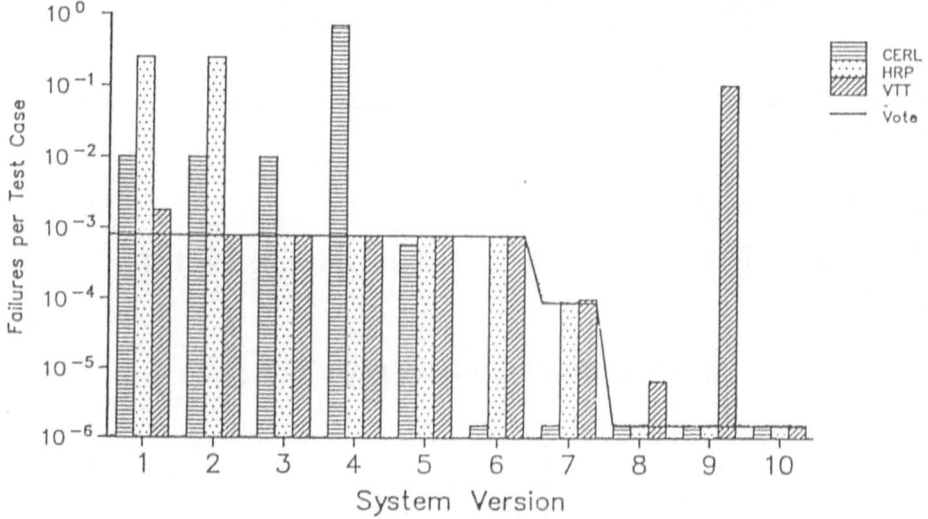

Fig. 7. System Failure Rate with Majority Vote

To check the sensitivity of the failure rate to the distribution of test values over the input domain, further back-to-back measurements were made on programs C1, H1, V1 using three different types of back-to-back test data (Table VI). The large variations in the failure rate highlight the fact that it is only possible to assess the failure rate of software for a specific mode of use. For example, the high failure rate of the HRP program was related to the processing of invalid input data. When a Gaussian distribution was used, the probability of out-of-range input values was reduced, so the failure rate was lower.

A comparison of two different test sequences with the same uniform random distribution produced very similar failure rates. This suggests that a probabilistic input distribution may be an adequate way of representing the mode

Table VI. Variation of the Failure Rate with Test Type

Test Type	Cases	Failure Rate		
		CEGB	HRP	VTT
Systematic	2472	.043	.53	.023
Uniform	65000	.010	.25	.002
Gaussian	65000	.002	.069	.088

of use. However more measurements would be needed to validate this hypothesis.

10.3 The Cost of Diversity

The amount of effort expended on the mainstream project activities is shown in Table VII.

Table VII. Effort Expended on Project Activities

Activity	Man-Hours Expended			
	SRD	CEGB	HRP	VTT
Proj. Management	546	213	246	170
Customer spec.	280			
Manufac. spec.+ QA		441	138	43
Design + QA		200	185	110
Code Production		137	292	155
Acceptance	86	82	64	45
Back-to-back tests	264	50	180	119
MOTH development	72	-	120	-
Test data dev.	60	-	238	-
Total	1308	1123	1463	642

The majority of the project management effort was devoted to the organisation of the experiment and should not be considered part of the normal software development process. The effort expended by the three teams on the specification, design, code production and acceptance phases was 1892 hours, so the average effort for one team would be 631 hours. However the cost of the customer requirement specification and acceptance test design was essentially independent of the number of versions and required a total of 366 hours of effort. The comparative cost of a single development would therefore be 997 hours compared with 2258 hours for three-fold diversity.

If the back-to-back testing phase is regarded as part of the development process rather than in-service operation, then there are no comparable figures for the additional verification and testing required to detect the same set of residual faults for a single program development. We intend to examine the performance of alternative fault detecting methods in a later experiment.

11. Impact of Specification Techniques

11.1 Experimental Difficulties

The reactor trip system was not complex enough to exercise more than a small part of the X specification facilities. In addition, although CEGB did not use a specification language, the customer requirements were analysed in some depth to locate potential problems. CEGB also tried to raise the customer's requirement specification to a higher level of abstraction using state transition diagrams. In this respect the experimental requirements were not fulfilled because CEGB's approach to specification was not informal enough.

11.2 The X Specification Language

The X specification language enabled the function and organization of the trip software to be specified in detail, describing all the component parts and their interactions. Using the computer-aided tool SPEX a variety of consistency checks and analyses could be made. The X specification was precise, unambiguous and easy to interpret. Only one fault in the X specification persisted into the back-to-back test phase and this was due the use of free text (which is permitted in X) rather than a rigorous mathematical definition. HRP also used X as a software design aid and, as such, it was quite successful. It was easy to translate the X specification into a program.

11.3 CEGB Specification Approach

The customer's requirement specification described the primary and secondary trip logic using terms similar to a conventional sequential programming language (e.g. if condition then ... else ...). However the order of the conditions differed between the main body of the specification and a flowchart example. This lead to a specification inconsistency. In the CEGB manufacturer specification, the trip logic was re-formulated using Boolean logic and state machine diagrams. As a consequence a number of apparently unnecessary states were found in the customer's primary and secondary trip logic. Fig. 8 shows the CEGB specification of the secondary trip logic, compared with that derived from the PODSPEC example. These problems were reported but never properly resolved. The apparently unnecessary states were removed from the CEGB specification. This gave rise to test "failures" in the CEGB version of the trip logic during acceptance testing.

Discrepancies were also detected in the secondary trip logic during back-to-back testing. However, on analysis it was agreed that the initial customer specification was erroneous, resulting in a common secondary trip logic fault in the HRP and VTT programs.

CEGB

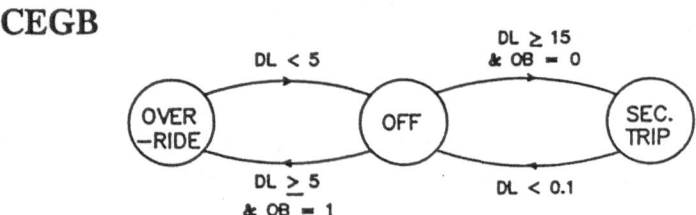

SRD

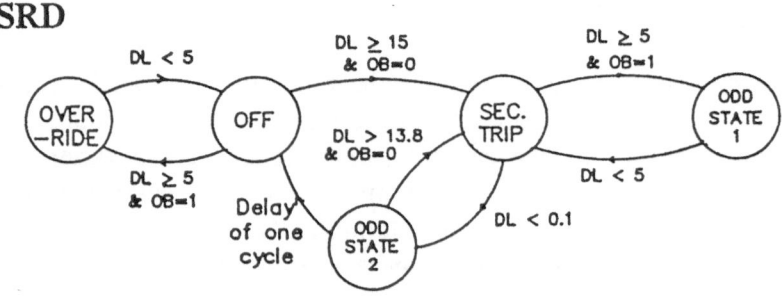

Fig. 8. Comparison of the Secondary Trip Logic

11.4 Comparison of Methods

The two specification methods appear to be complementary. The reformulation of the customer requirements in more abstract terms showed up potential faults in the customers statement of requirements, while the X specification helped to ensure that the software specification was internally consistent and complete. The manually-produced CEGB specification did not suffer from any major internal inconsistencies or omissions, but these could have occurred in a more complex specification.

12. Analysis of Programming Language

12.1 Number of Coding Faults

More than a hundred coding faults were reported by the three teams during the code production phase, but they were detected either by local testing or by acceptance testing. No coding faults were detected in the subsequent back-to- back tests. The apparent lack of residual coding faults can probably be attributed to the lack of complexity in the system, enabling fairly exhaustive testing and inspection to be carried out. In general such testing may not always be possible. In these circumstances it is probable that the number of residual coding faults would be related to the number of errors made during code production, and so the numbers of faults created in the code production phase have been examined for each of the three teams (Table VIII).

Table VIII. Coding Faults for Fortran and Assembly Language Programs

Team	Language	Faults	Code Lines/Fault
CEGB	Fortran	25	34
HRP	Assembly	87	22
VTT	Fortran	5*	95

* VTT did not record faults made during code production

Since VTT did not record faults within the code production phase, only the faults in the CEGB and HRP programs can be compared. From this comparison it would seem that the use of a high level language reduced the number of coding faults by at least a factor of 3. Since the assembly language program was about three times longer than the Fortran programs, this was consistent with observations from other sources [Lipow 1982] that the number of faults per line is similar regardless of the programming language.

12.2 Programming Effort

For the purposes of the analysis shown in Table IX, programming was defined as coding plus development testing (both module and integration). Other activities such as manufacturer's specification and local test harness production were not included.

Table IX. Man-hours of Programming Effort

Activity	CEGB (Fortran)	HRP (Assembly)	VTT (Fortran)
Coding	58	150	74
Code QA	4	46	36
Dev. Testing	75	96	45
Total	137	292	155

It would appear that the HRP assembly language program took about twice as long to write as the equivalent FORTRAN programs.

12.3 Program Length

The CEGB and VTT Fortran programs were considerably shorter than the assembly language program produced by HRP. The characteristics of the programs produced by each team are summarized in Table X.

Table X. Program Characteristics

Team	Modules	Code Lines	Total Lines
CEGB	35	859 Fortran	3235
HRP	27	1906 Assembly	3407
VTT	17	477 Fortran	797

12.4 Performance

The performance of each program, running on the NORD computer is shown in Table XI.

Table XI. Run-Time Performance of Each Program

Mode	Response (msecs)		
	CEGB	HRP	VTT
Initialize	294	266	359
Calibrate	292	258	354
Init.&Cal.	339	266	359
Normal	246	32	87

All the programs responded within the specified time of 500 milliseconds. The CEGB program response times are relatively constant in all modes. This was probably due to a design decision to avoid as much conditional coding as possible, so that a large proportion of the initialization code was executed on every program cycle. In normal operation, the HRP assembly language program was nearly three times as fast as the equivalent VTT program which had the same reactor power algorithm and initialization strategy.

12.5 Faults in Support Tools

No faults were discovered in the VAX VMS Fortran compilers used by CEGB and VTT for local development. One commonly expected declaration check was absent from the pre-release NORD assembler used by HRP. In addition two faults were discovered in a pre-release version of the NORD Fortran compiler used by CEGB during re-acceptance.

13. Analysis of the Software Development

13.1 Faults Reported

The faults reported by each team are summarized in Table XII.

The number of distinct faults in the SRD specification is difficult to determine, as the fault reports from each team tended to overlap, but not precisely. Around 90 distinct faults were attributed to the customer's

Table XII. Faults Reported by Each Team

Faults Reported in:	CEGB	HRP	VTT
Customer specification	68	43	11
Manufacturer specification	53	42	56
Design specification	19	26	3
Coding	26	87	5
Acceptance modification	6	0	0
Back-to-back modification	1	0	1
Test data	1	0	0
Support software	2	0	0
Total	176	198	76

requirement specification.

Fig. 9 shows the persistence of faults between the phase of creation and the phase of detection. It can be seen that the most persistent faults originated in the customer's requirement specification.

13.2 Detection of Faults During Development

The effectiveness of the different methods employed to detect faults during development is shown in Table XIII.

The HRP team performed the greatest number of inspections, and this is reflected in the higher proportion of faults detected by this method. The CEGB team noted that proper inspections required considerable initial preparation and this activity should have been incorporated into the development schedule. It would also have been desirable to have had an independent inspection team available from the beginning of the project. This would have minimized the need to provide and understand all the background information required for an inspection.

Table XIV shows the proportion of known faults detected within the phase where they were created. The proportion detected is quite high. This can probably be attributed to the quality assurance programme employed during development.

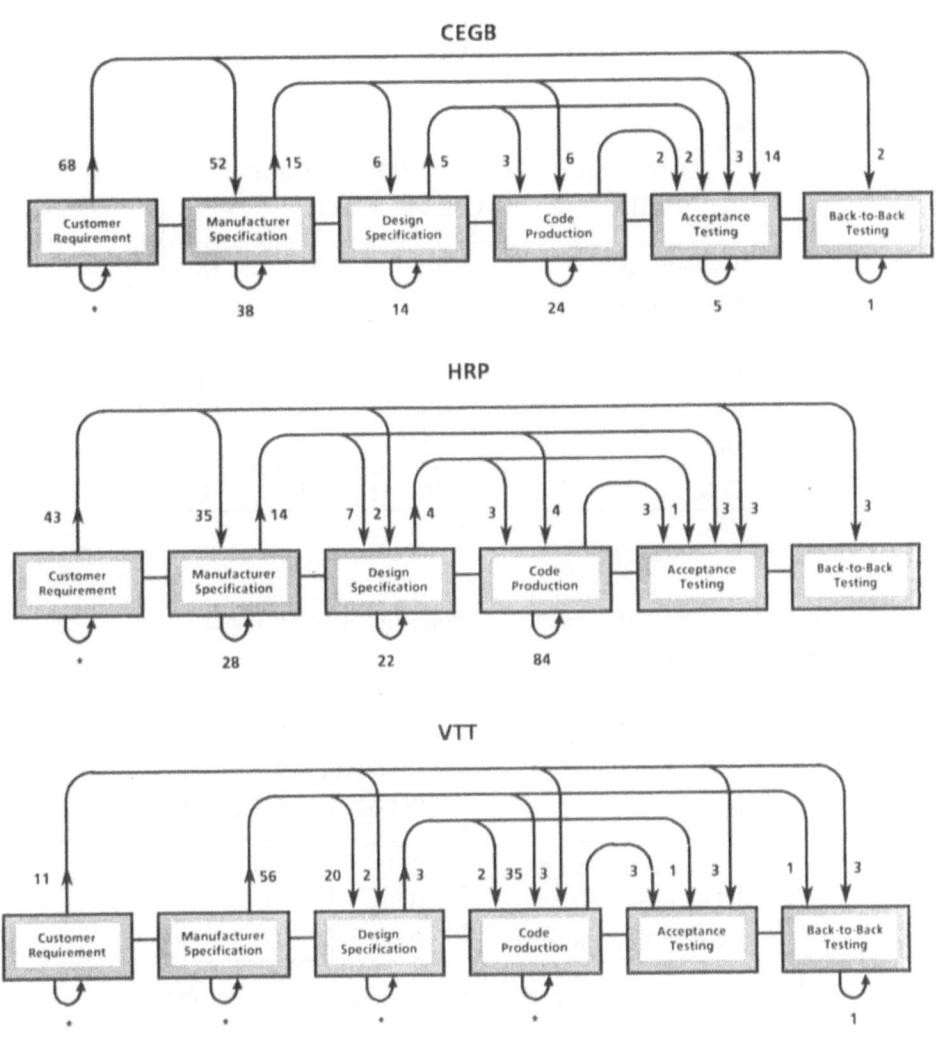

Fig. 9. Fault Persistence Between Phases

13.3 Post-Development Fault Detection

The numbers of faults detected by the acceptance and back-to-back tests are shown in Table XV.

For comparison, the same test data and procedures were used to locate 30 seeded faults in the "golden" version of the HRP program. The seeded faults were created manually, and inserted into the following types of code:

Table XIII. Fault Detection Methods

Detection Method	Faults Found		
	CEGB	**HRP**	**VTT**
Inspection	26	101	28
Walkthrough	14	8	1
Individual checks	81	19	27
Translation	4	33	1
Test	47	25	17
Other	4	12	-
Total	176	198	76

Table XIV. Fault Detection Efficiency per Phase

Phase	Fraction of Faults Detected *	
	CEGB	**HRP**
Manufacturer spec.	0.72	0.67
Design	0.74	0.85
Code Production	0.92	0.97

* VTT did not record faults occurring within a phase

- computation of an expression,
- branch conditions,
- constants definitions,
- initialization.

Faults were detected by comparing the outputs of the HRP program with the outputs of the "golden" CEGB and VTT programs. The test data was applied in the normal back-to-back test sequence. When a fault was detected, the program was corrected and the tests were repeated from beginning. Table XVI shows that 25 of the faults were detected by the test data.

An analysis of the remaining 5 faults showed that they could not have been

Table XV. Faults Detected During the Test Sequence

Test Data	Known Faults Remaining	Faults Detected
Acceptance	57	48
B-B Systematic	9	4
B-B Uniform Random	5	3
B-B Rectangular/Boundary	2	2
B-B Gaussian Random	0	0
B-B Gaussian/Boundary	0	0

Table XVI. Seeded Faults Detected by the Development Test Sequence

Test Data	Seeded Faults Remaining	Faults Detected
Acceptance	25	22
B-B Systematic	3	1
B-B Uniform Rand.	2	2

found by testing. One fault, for example, was in defensive code which should never be executed, while another fault affected a constant that was not used.

Both sets of results indicate that the acceptance test data was quite efficient, finding around 85% of the known faults that were capable of detection. The back-to-back tests successfully detected the remaining seeded faults, giving some confidence in the coverage of the test data.

Direct comparisons between the various types of test are difficult to make because the faults were removed as they were detected, leaving fewer faults for each successive test sequence. It would be interesting to examine whether back- to-back testing using random data is more effective at locating faults than the systematic tests. Random testing enables a large number of tests to be applied, but it cannot detect a fault which is common to all programs. However, this form of testing would obviously be much easier to design and apply, so reducing the cost of testing. An analysis of the relative effectiveness of the tests will require further experiments where the fault detection capability and coverage of the individual tests are measured for all three programs.

13.4 Project Organization

Most of the problems encountered during the project were related to difficulties in communication or information processing. These problems had to be solved as the project progressed. On the basis of this experience, we would recommend that in any similar project:

- The organization, documentation, and terminology should be defined and understood by all parties at the outset.
- Good communication channels should be established between all parties. Frequent meetings should be held and telex, telefax and electronic mail should be employed to reduce response times.
- Computers should be used for project support. All documentation, project management information and experimental data should be stored on a computer in a form which can be easily maintained and analysed.

14. Conclusions

It is not possible to draw general conclusions from the results of a single experiment, but within the context of this experiment, we present the following conclusions.

Diverse implementation

The use of diversity was beneficial in the following respects:

- Diverse implementation was effective in reducing the failure rate. A trip system composed of three diverse acceptance-tested programs combined with majority voting had a significantly lower failure rate than the failure rate of any individual program, when tested with random input data.
- Diverse implementation also provided an economic way of carrying out a large number of tests. The diverse programs lent themselves to automatic testing by comparing of their responses to randomly generated input values.

However, some disadvantages were also observed:

- Diverse implementation did not eliminate all common faults. Out of the seven residual faults, two faults were common.

- The cost of diverse implementation was at least twice that of a single program development.

Specification

- The computer-based formal specification language "X" was useful both in specification and design. However, the example application was insufficiently complex for a complete assessment to be made.

- Most of the residual faults were caused by the requirement specification. It is likely that these could have been reduced if better methods of acquiring, analysing and specifying the requirement had been used.

Comparison of languages

- Assembly language programming required about twice as much coding effort and generated three times as many coding faults as Fortran programming.

Software development

- Careful use of recommended development and testing methods eliminated design and coding faults soon after they arose, and no implementation faults were detected in the final programs.

15. Future Work

The PODS project raised several questions that could not be tackled within the original time-scales. For example, we do not know if the residual faults (and possibly others) could have been found by more analysis or different testing approaches, and we do not know if alternative approaches could have been more cost-effective.

A follow-up project, called STEM, has been launched to examine this area in more detail. STEM is an acronym for Software Test and Evaluation Methods project and it will make use of the existing versions of the PODS trip software, which contain known faults. The software will be checked using a variety of static and dynamic techniques in order to assess their relative effectiveness and efficiency at detecting faults. These techniques will include:

- manual inspection

- code analysis tools

- test coverage measurement

- test strategies using realistic, random, and systematic test data.

The effort required for each technique will also be measured so that realistic comparisons can be made between alternative methods.

STEM is also examining the existing faults within the PODS programs to measure the properties of software faults that are relevant to software reliability modelling. The PODS programs are a useful experimental vehicle since they possess internal states which vary between execution cycles, a common feature of real-time software. The overall model for software failures is shown schematically in Fig. 10. We expect that internal state effects and different input distributions will both have an impact on the observed failure rate, as our preliminary results have indicated (Table VI). The main aspects that will be investigated are:

- the correlation between independent faults.

- the distribution of failure probabilities for the population of faults.

- the validity of bug-seeding as an estimator of residual faults.

- the relationship between the input distribution and the failure rate.

- the predictive ability of some of the existing software reliability models.

The results of this study should be reported in 1987.

The PODS experiment also confirmed that the initial functional specification is a very significant source of persistent, common mode faults. CEGB has mounted a project to study the specification of a safety interlocking system using mathematically formal specification and analysis techniques.

Acknowledgements

The author wishes to acknowledge:

A. Ball (SRD),
M. Barnes (SRD),
P. Humphreys (SRD),
D. Esp (CEGB),
P. Rutter (CEGB),
G. Dahll (HRP),

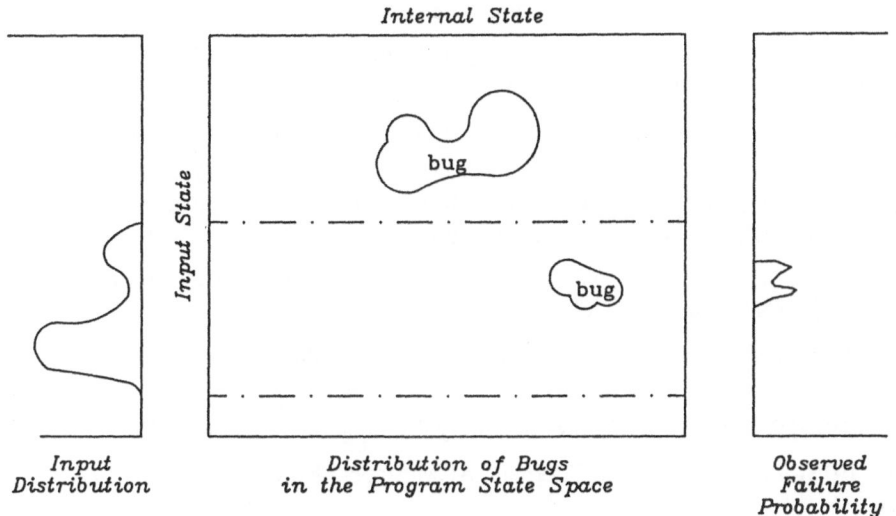

Fig. 10. Schematic Diagram of Fault Activation

O. Hatlevoldt (HRP),
S. Yoshimura (NAIG/HRP),
J. Lahti (VTT), and
B. Bjarland (VTT)

for their contributions to the PODS project. The author also thanks the Halden Reactor Project and the Halden UK Control and Instrumentation Liaison Committee for their support.

This contribution is based on a paper published in the September 1986 issue of the IEEE Transactions on Software Engineering [Bishop 1986].

References

[Alford 1973] R. W. Alford, "A Requirement Engineering Methodology for Real-Time Processing Environments", IEEE Trans. on Software Engineering, Vol. SE-3, No. 1, January 1973.

[Avižienis 1975] A. Avižienis, "Fault-Tolerance and Fault-Intolerance, Complementary Approaches to Reliable Computing", Proc. 1975 Int. Conf. Reliable Software, Los Angeles, 1975.

[Barnes 1985] M. Barnes et al, "PODS (The Project on Diverse Software)", OECD Halden Reactor Report, HPR-323, 1985.

[Bishop 1986] P. G. Bishop et al, "PODS - A Project on Diverse Software", IEEE Trans.

on Software Engineering, Vol. SE-12, No. 9, pp. 929-940.

[Dahll 1983] G. Dahll and J. Lahti, "The Specification System X-SPEX", IFAC Conference "Safety of Computer Control Systems", Cambridge, UK, pp. 111-118, 1983.

[EEA 1981] "Guide to the Quality Assurance of Software", Electronic Engineering Association, 1981.

[Fagan 1976] M. E. Fagan, "Design and Code Inspections to Reduce Errors in Program Development", IBM Systems Journal, No. 3, pp. 182-211, 1976.

[Lipow 1982] M. Lipow, "Number of Faults per Line of Code", IEEE Trans. on Software Engineering, Vol. SE-8, No. 4, July 1982.

[Minsky 1967] M. L. Minsky, Computation, Finite and Infinite Machines, Prentice Hall, 1967.

[Myers 1976] G. J. Myers, Software Reliability Principles and Practices, Wiley, 1976.

[Nassi 1973] I. Nassi and B. Shneiderman, ACM Sigplan Notices, 8 (August 1973) 8, pp. 12-26.

[Yourdan 1975] E. Yourdon and L. Constantine, Structured Design, Yourdon Inc, 1975.

4

Flight Applications

This chapter gives an overview on some applications in the aircraft industries. The paper by Traverse presents some detail on the use of Software Diversity by Aerospatiale. Besides in the Airbus, software diversity is also applied in other aircraft environments.

At the Workshop, two further presentations were given, by Wright from GEC Avionics and Yount from Sperry. Since their contributions could not be included in this volume, a brief report on their talks is given here.

Nigel Wright gave a talk on the use of software diversity in the Airbus A310 [Wright 1986]. The slat/flap control system of the A310 consists of two functionally identical computers with diverse hardware (see Fig. 1). They are manufactured by the Flight Controls Division of GEC Avionics Ltd. in Rochester, UK. Within each computer, two diverse programs are executed whose results are compared via an AND-logic (see Fig. 2). For the synchronization of the two versions, the AND-logic has a 350 ms window.

The diversity architecture was chosen because
- high integrity requirements exist,
- the availability is a less stringent requirement,
- the extent of the task and the ability to use avionic grade microprocessors, and
- the certification risk.

The diversity was ensured by independent design teams, use of diverse hardware and separate host facilities for the software design environment (see Fig. 3). Besides applying very rigorous testing before the release of the system, there is a continuous real time test during the use of the system.

The disadvantages like doubled software size, doubled design documentation, extra host system equipment and lower availability were overcome by the advantages like higher safety, lower certification risk, clarification of the specification, configuration control and continuous real time test.

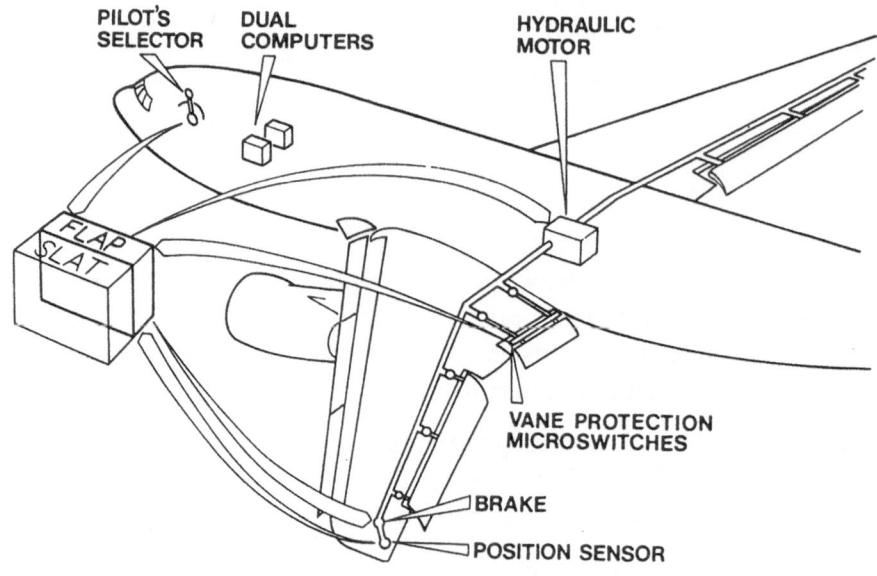

Fig. 1. A310 System Configuration (Flap only illustrated) [Wright 1986]

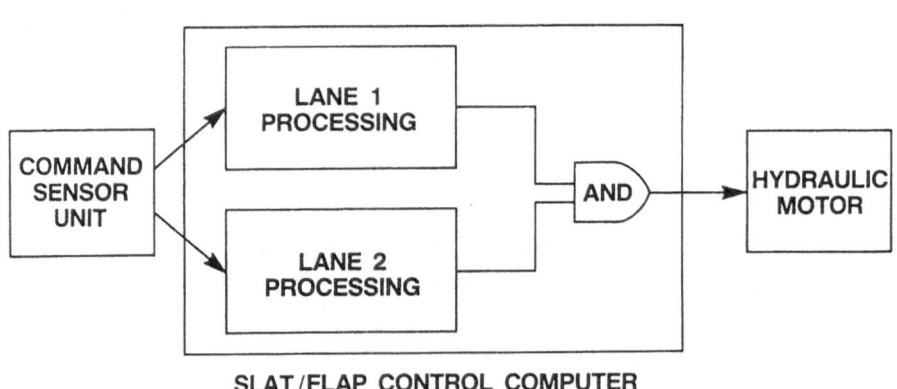

Fig. 2. Slat/Flap Control System Principle [Wright 1986]

Furthermore, the diverse redundancy was preferred to the similar redundancy, e. g. because the similar redundancy has a higher risk of an undetected design error adversely affecting the system safety.

The use of software diversity made the application of high level languages possible, since two different languages were used and therefore compiler validation was considered unnecessary. The use of high level languages

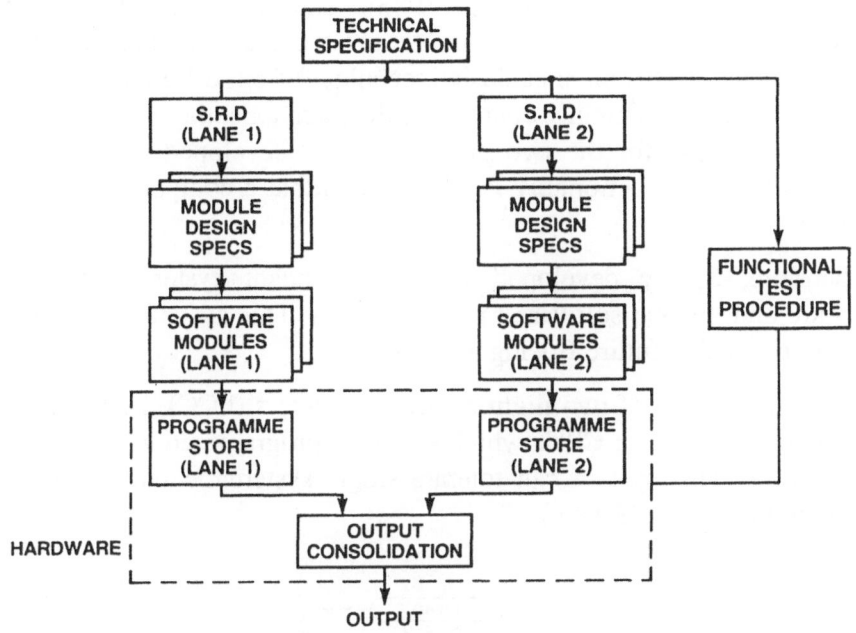

Fig. 3. Software Development Process [Wright 1986]

increased the software productivity.

As a further fault tolerance means, the outputs of the two programs are compared not only with each other, but also with an estimate generated from the previous cycle. The best of these data is output to the actuators.

The system was certified in March 1983. Since its introduction to airline service, only one revision to the software was necessary, mainly due to performance improvements. No erroneous deployment of the surfaces was reported, and the reliability of the computers (in the sense of MTBF) is exceeding the expectations.

Several aspects of this approach are also described in [Martin 1981, Martin 1982, Hills 1983, Hills 1985].

The next presentation was given by Larry Yount on the projects at Sperry Corporation (now Honeywell - Sperry Commercial Flight Systems Division) on software diversity [Yount 1986]. He described the system SP-300 which is in use in several Boeing aircraft and a new system for future aircraft, which is in development.

The autopilot flight director system SP-300 for the Boeing 737-300 aircraft was developed by Sperry [Williams 1983]. The SP-300 AFDS consists of two redundant computers. Each computer again contains two diverse processors (microchip versus bit slice design technique) from different manufacturers. One processor is mainly controlling the pitch axis, the other one the roll axis. The software for the two processors was developed by independent teams. Part of the functionality is identical, so that a comparison check can be made.

Furthermore, Sperry developed a dual redundant system with diverse hardware and diverse software [Yount 1985b]. This system design is supposed to be used in future Boeing aircraft.

The system consists of two flight control computers (FCC), each containing three diverse redundant CPUs which again are programmed in a diverse way (see Fig. 4). This system can tolerate single systematic faults within one design unit (CPU + software).

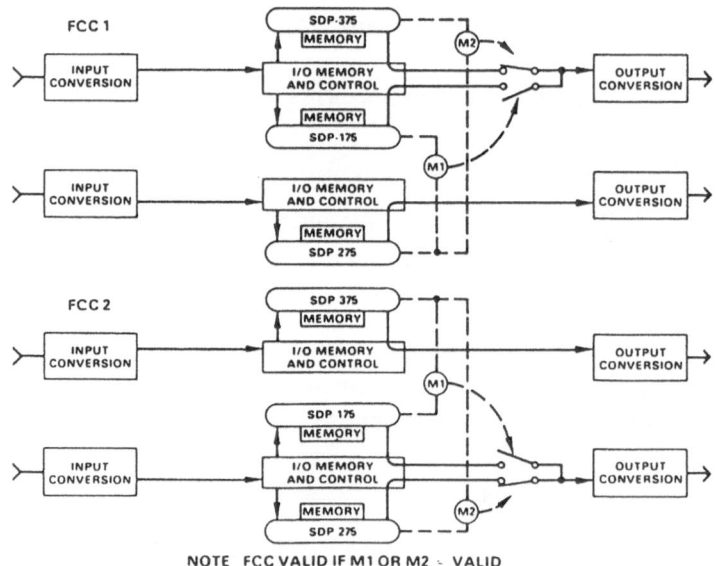

Fig. 4. Sperry's Dual Architecture [Yount 1986]

The development process by which Sperry tries to reduce the amount of generic faults is shown in Fig. 5.

The main advantage of diversity is seen to be the gain in overall system reliability.

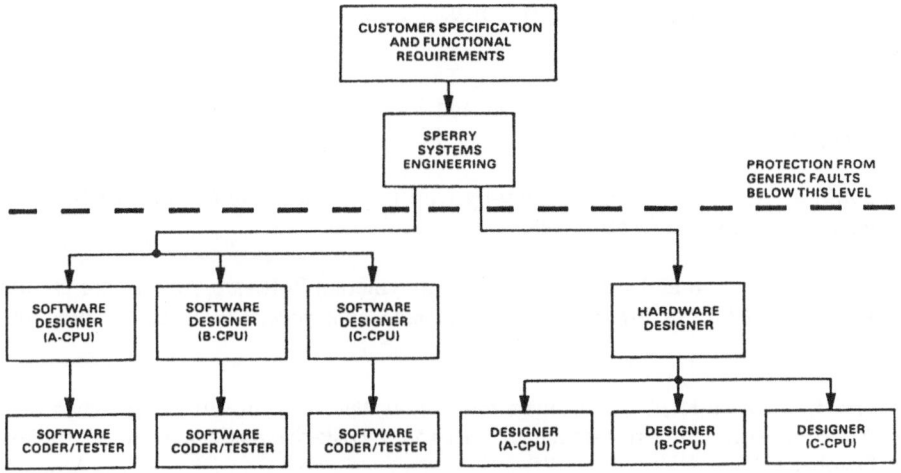

Fig. 5. Protection from Generic Software and Processor Faults [Yount 1986]

Further references to the work by Sperry are [Yount 1984, Yount 1985a].

The Boeing 757/767 is equipped with a yaw damper making use of two version programming [Yount 1985a].

Further references to the use of diversity in the airplane industry includes [Garman 1981, Hack 1983, Hitt 1984, Hofer 1983].

Besides the use of diversity for on-line purposes the parallel development of a second system for testing purposes only is also reported [Stocker 1983].

Within the papers from the aircraft industry, the use of the term *dissimilar software* is diversity.

Aircraft flight control systems developed by Rockwell Collins use dual diverse microprocessors as well as software diversity. They are in use in fail-safe systems as well as in fail-passive systems [Turner 1987].

This wide variety of use of software diversity in a safety critical environment demonstrates the applicability of this technique and a strong belief by these industries that this technique is a cost-effective and necessary technique to increase the dependability of the system.

References

[Garman 1981]   J. R. Garman, "The 'Bug' Heard 'Round the World," *ACM Sigsoft SEN,* Vol. 6, No. 5, October 1981, pp. 3-10.

[Hack 1983]   J. P. Hack, "Digitale Elektronik in Verkehrsflugzeugen (Digital Electronic in Airplanes - in German)," in *DGLR-Symposium,* Köln, Germany: 25-26 October
1983.

[Hills 1983]   A. D. Hills, "A 310 Slat and Flap Control System Management and
Experience," in *Proc. 5th DASC,* November 1983.

[Hills 1985]   A. D. Hills, "Digital Fly-by-wire Experience," in *Nato AGARD Conf.,*
Edmunds AFB, CA, USA: October 1985.

[Hitt 1984]   E. F. Hitt and J. J. Webb, "A Fault-Tolerant Software Strategy for Digital
Systems," *AIAA/IEEE 6th Digital Avionics Systems Conference,* 3-6 December 1984, pp.
211-216.

[Hofer 1983]   H. Hofer, "Erfahrungen mit Flight Standard Software (Experience with
Flight Standard Software - in German)," in *Proc. DGLR-Symposium,* Köln, Germany:
25-26 October 1983.

[Martin 1981]   D. J. Martin, "Dissimilar Redundancy for Fly-by-wire Secondary
Flight Controls," in *Proc. Advanced Flight Controls Symposium,* Colorado Springs, CO,
USA: 1981.

[Martin 1982]   D. J. Martin, "Dissimilar Software in High Integrity Applications in
Flight Controls," in *Proc. AGARD Symp. on Software Avionics, CPP-330,* The Hague,
The Netherlands: September 1982, pp. 36.1-36.13.

[Stocker 1983]   J. Stocker and J. Rauch, "CSTS: Ein Software-Testsystem für den
Tornado-Autopiloten (CSTS : A Cross Software Test System for the Tornado Autopilot -
in German)," in *Proc. DGLR-Symposium,* Köln, Germany: 25-26 October 1983.

[Turner 1987]   D. B. Turner, R. D. Burns, and H. Hecht, "Designing Micro-Based
Systems for Fail-Safe Travel," *IEEE Spectrum,* Vol. 24, No. 2, February 1987, pp. 58-
63.

[Williams 1983]   J. F. Williams, L. J. Yount, and J. B. Flannigan, "Advanced
Autopilot-Flight Director System Computer Architecture for Boeing 737-300 Aircraft,"
in *Proc. Fifth Digital Avionics Systems Conference,* Seattle, Washington, USA: 30
October - 3 November 1983.

[Wright 1986]   N. C. J. Wright, "Dissimilar Software," in *Workshop 'Design Diversity in Action',* Baden, Austria: 27-28 June 1986.

[Yount 1984]   L. J. Yount, "Architectural Solutions to Safety Problems of Digital
Flight-Critical Systems for Commercial Transports," in *Proc. of the AIAA/IEEE 6th
Digital Avionics Systems Conf.,* Baltimore, MD, USA: 3-6 December 1984, pp. 28-35.

[Yount 1985a]   L. J. Yount, "Generic Fault-Tolerance Techniques for Critical Avionics Systems," in *Proc. AIAA Guidance and Control Conference,* Snowmass, CO, USA:
June 1985.

[Yount 1985b]   L. J. Yount, K. A. Liebel, and B. H. Hill, "Fault Effect Protection and Partitioning for Fly-by-Wire and Fly-by-Light Avionics Systems," in *Proc. AIAA/ACM/NASA/IEEE Computers in Aerospace V Conference,* Long Beach, CA, USA: October 1985, pp. 275-284.

[Yount 1986]   L. J. Yount, "Use of Diversity in Boeing Airplanes," in *Workshop 'Design Diversity in Action',* Baden, Austria: 27-28 June 1986.

AIRBUS and ATR
System Architecture and Specification

Pascal Traverse
AEROSPATIALE
316, Route de Bayonne
31 060 Toulouse Cedex
France

1. Introduction

Design diversity is widely used on board of the aircraft for which AEROSPATIALE has the design responsibility. Indeed, all the aircraft currently in production, and all the aircraft designed since the end of the 70's are using dissimilar software. The current (mid-1986) number of such aircraft is 120 AIRBUS, and 20 ATR.42. A lot of different functions are computerized with dissimilar software. The principal ones are the automatic and electric flight control systems ("automatic pilot" and "Fly by Wire"), and the flight instruments.

It is needed to cope with a large set of different faults to reach the dependability objectives of a computer-based avionics system. Different techniques are used, design diversity being one of them. For more information on these techniques, see [Rouquet 1986].

Design diversity is particularly targetted to tolerate design faults in order to reach safety and reliability objectives. Side effects are the tolerance of some external physical faults, and an efficient help to remove software errors (doing a so called "back-to-back" test, i.e. running all the different programs in parallel, with a comparison of the output).

This paper will deal basically with design diversity. With this focus, the global architecture of the on board computing system is presented, with the specification process, and the accrued experience. An annex lists most of the means that are used to have a high level of diversity.

The terms on dependability and design diversity used in this paper are intended to be coherent with the taxonomy published in [Avižienis 1986].

2. System Architecture

The dependability objectives for functions like:

-   Fly by Wire
-   Automatic flight control
-   Flight instruments

are twofold:

-   *safety*: a failed computer must not be able to send a wrong output. This leads to a *safe-shutdown* requirement.
-   *reliability*: these functions must be continuously available, and have therefore to be *fault-tolerant*.

2.1 Design Fault Detection and Tolerance

The safety requirement is fulfilled using as a building-block a pair of computation channels. Each channel contains (at least) one CPU, input/output devices, and its own program. Typically, only one of the two channels is controlling the output of the equipment. This channel is called the "control" or "command" channel and the other the "monitor" one (see Fig. 1).

In fact, each channel is monitoring the other, and is able to shut-down the whole computer. This shut-down is designed to be safe. Design diversity is used in the sense that the two channels have different programs.

Therefore, if one of the programs is faulty, this will be detected by comparing the output of the two programs. If the discrepancy is above some threshold and persists after a few retries, the computer shuts itself down.

Comparison between the two channels is not the only error detection mechanism. Acceptance and exception testing, and watchdogs are also used. The "command" processor, and the "monitor" processor are often of the same type. Because of the two different programs, it is very likely that a

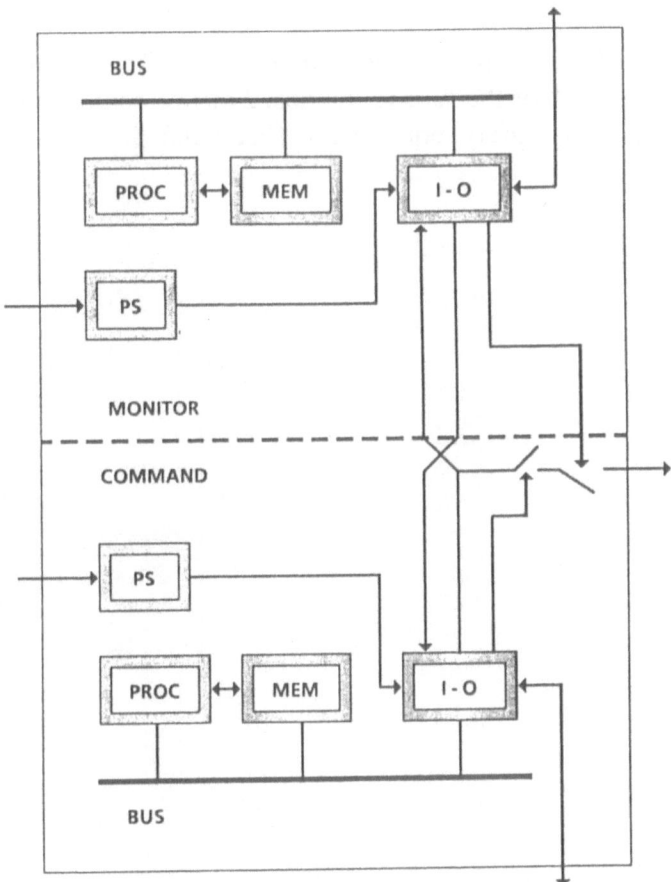

Fig. 1. Basic Layout of Computer

microprocessor design error would behave as a software error of one of the programs. More, as the used processors are hardened versions of commercial ones, it can be argued that they are extensively tested.

The tolerance to design faults is achieved using a (electro) mechanical back-up, or a design fault-tolerant computing system, as for the "fly-by-wire" A320.

We will use the two examples of the roll control of the A310, and of the pitch control of the A320, to illustrate the two design fault-tolerance approaches.

2.2 A310 Roll Control

The implementation of this function is depicted on Fig. 2. Four computers are involved and each of the four is composed of two computation channels, as defined to reach the safety requirements (Paragraph 2.1).

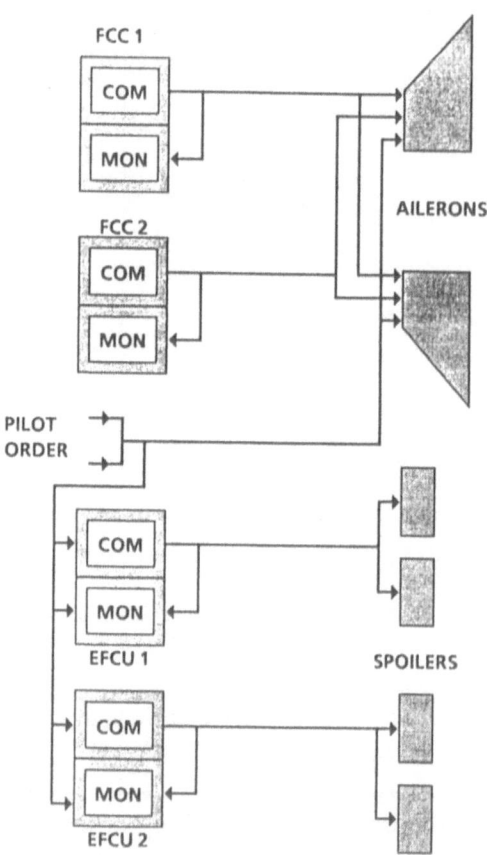

Fig. 2. A310 Roll Control

The aircraft is controlled on the roll axis using a pair of ailerons and 5 pairs of spoilers. The aircraft is controlled either in a manual mode, or in an automatic one. The automatic flight control system is composed of two "Flight Control Computers" (FCC1 & FCC2). Only one of them is active at a time, the other one being a spare. If both computers are lost, the aircraft is manually controlled. Therefore, the loss of the automatic flight control system is not dangerous for the aircraft (except during a short period of an automatic landing in bad weather conditions).

Computers are also involved in the manual control mode. Two "Electrical Flight Control Units" (EFCU1 & EFCU2) are used to control the spoilers. If both of these computers are lost, the pilot can still control the aircraft using the ailerons, with a reduced authority, as the spoilers are no more available.

2.3 A320 Pitch Control

The A320 is, as far as we know, the first aircraft using a computing system able to tolerate design faults. Diversity is such that two different types of computers are used, one (named ELAC) manufactured by THOMSON-CSF, based on 68000 type processors, the other (named SEC) by SFNA and AEROSPATIALE with 80186 processors. Of course, each computer is built with two computation channels and two different programs. Therefore, four different programs are used.

Apart from these computers, a limited mechanical back-up can be used. The design objective is for the computing system to be sufficiently reliable, in order not to use this mechanical back-up.

Four computers are used to control the aircraft on the pitch axis: ELAC1, ELAC 2, SEC1, SEC2 (see Fig. 3).

At each time, only one is needed to have the full authority. The computer in charge of the pitch axis is periodically sending "I am alive" like messages to the other computers. If this computer fails, it shuts itself down, and this will be detected by the other computers. According to a predefined priority, one of them will take the control of the pitch axis.

Something that can happen in case of a design fault is for one type of computers to shut itself down. In this case, one computer of the other type has to take the control. The whole system is thus able to tolerate design faults.

The Fly by Wire system is intended to improve the safety of the aircraft. It will provide a protection against windshear, a protection of the flight envelope, and an alleviation of the burden of the pilots.

More details about the A320 Fly by Wire system can be found in [Ziegler 1984, Corps 1985].

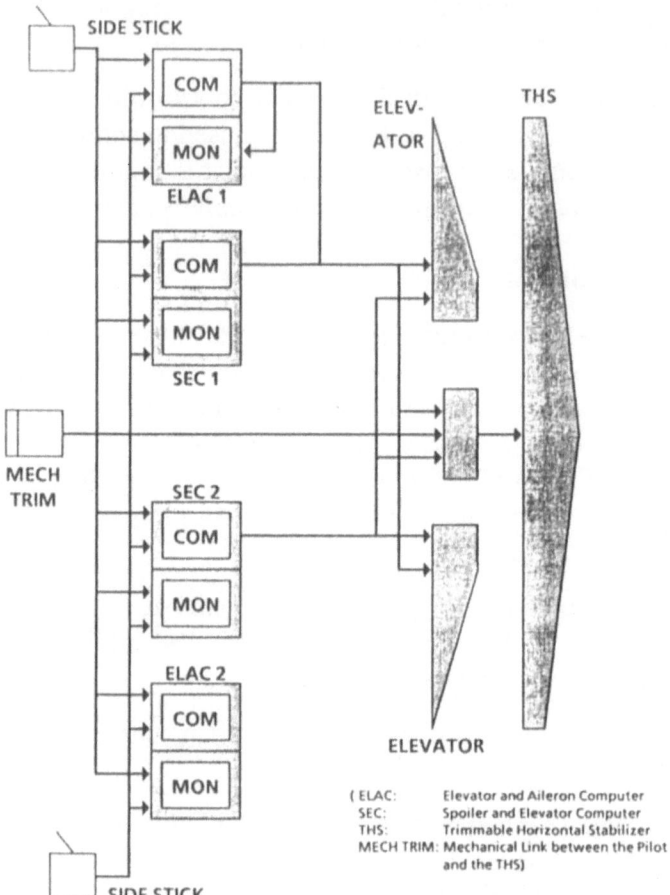

Fig. 3. A320 Pitch Control

3. Specification Process

The specification process is based on some kind of formal specification and of rapid prototyping. The formal aspect is based on a high degree of modularity and of the use of logic symbols (AND, OR), and transfer operators (filters, integrators). The operational logic and the control laws of an equipment are specified in a graphical form, using these symbols. The rapid prototyping aspect comes from the fact that either the specifications are executable (operational logic), or simulated (control laws), and from the fact that the specification changes are normal during the environmental testing in iron birds (i.e. almost a grounded aircraft), and even after the first flights. In fact, apart from an unexpected event, the only expected specification changes are parameter tuning after the first flight.

The operational logic part of the specification is automatically programmed. When dissimilar software is used, their coding algorithms are different, as are their logic operator coding libraries.

A criticality level is a part of a software specification. To each level corresponds a software development and testing plan, as defined in [DO178A 1985]. Because of the use of dissimilar software, it is allowed by the Airworthiness Authorities to downgrade the criticality level, in a few cases. This is not allowed for the Fly by Wire A320, and for any Automatic Flight Control System, if intended to land in bad weather condition.

There is not standard design diversity specification. The general practice is to require:

- different software design teams
- different languages
- the equipment manufacturer has to define programming rules in order to increase the diversity between the control and monitor programs.

Most of the rules used to enforce diversity are listed in the annex.

4. Experience

Our experience with safety computer is now noticeable (see Table 1).

Table 1. Experience

AIRCRAFT	A300 FFCC	A310, A300-600	ATR.42
In Service Year	1982	1983	1985
Number of Aircraft	15	106	20
Flight Hours	100 000	450 000	25 000
Computer Types [2]	3	7 [1]	2
On Line Computers [3]	4	11 [1]	3

[1] One more on extended range aircraft
[2] EFCU, FCC,...
[3] Computers that are on board and normally used; hot and cold spares are not counted

Numerous functions are implemented with dissimilar software and this involves numerous equipment manufacturers. Each of the computers mentioned in Table 2 is of the type depicted by Fig. 1, with two diverse programs. Our record is satisfactory. No aircraft crashed or even came close to this situation.

Table 2. Number of Types of Computers Using Diverse Software

Aircraft	A300 FCC	A.310 A300-600	A.320	ATR.42
Total	3	7 or 8 [1]	7	2
Auto Flight	3, SFENA	3, SFENA	2, SFENA	1, SPERRY
Flight Instrument		1, Thomson CSF	1, Thomson CSF	1, SPERRY
Fly by Wire [2]		1[3], GEC 1, Thomson CSF 1, AEROSPATIALE	1[3], GEC 1, Thomson CSF 1, SFENA/ AEROSPATIALE	
Others		1[1,3], SFENA	1, AEROSPATIALE	

[1] Extended range aircrafts only

[2] Almost complete for the A320, only partial for the others

[3] AEROSPATIALE has not the design responsibility

The quality of the software is good. For the computers mentioned in Fig. 2, either no software error has been reported (EFCU, 8 kW per program), or the only ones that have been detected are benign (FCC, 120 kW). As for related software error (an error in both versions of a software), not even one has been reported. Report mechanisms are the pilots and on-line error logging devices.

As these errors are benign, we rather live with them until it is decided to release a new software version (with more accurate maintenance function, new functionalities requested by airlines). As the approach for diversity specification and enforcement is successful on the A310, no major change has been done on the A320 on this point.

5. Conclusion

Design diversity will be used on our next aircraft: ATR.72, AIRBUS 330 and AIRBUS 340. At this time, it seems that having two different versions of a software had not been necessary in operation, due to the high quality of each program. On the other hand,

- design diversity is helping us to find some specification ambiguities, and to test the software, and
- we feel more confident having different programs rather than relying on a single one.

References

[Avižienis 1986] A. Avižienis and J. C. Laprie: Dependable Computing: From Concepts to Design Diversity. Proceedings of the IEEE Vol. 74, No. 5, May 1986, pp. 629-638.

[Corps 1985] S. G. Corps: A320 Flight Controls. Proceedings of the 29th Symposium of the"Society of Experimental Test Pilots", September 1985.

[DO178A 1985] Software Considerations in Airborne Systems and Equipment Certification, Radio Technical Commission for Aeronautics (R.T.C.A.), No. DO178A, March 1985. Also published by the European Organization for Civil Aviation Electronics (EUROCAE) No. ED-12.

[Rouquet 1986] J. C. Rouquet and P. J. Traverse: Safe and Reliable Computing on board of AIRBUS and ATR Aircraft. Proceedings of SAFECOMP '86, October 1986, SARLAT, France, pp. 93-97.

[Ziegler 1984] B. Ziegler and M. Durandeau: Flight Control System on Modern Civil Aircraft. Proceedings of the International Council of the Aeronautical Sciences (ICAS'84), Toulouse, France, September 1984.

Annex

Diversification means

The diversification means are tentatively classified. It has to be noted that all the computers are not using all the listed means, but only a sub-set. Basically the goal is to have two different software development processes, from the software specification to the programming tools, including the software designers. Diversity is enforced in two general cases: on alternatives (two possible algorithms, for example) to avoid both design teams to make the same choice, and when a point in the specification is felt to be complex. Examples of the later are trigonometric computation (use of polar coordinates in one channel, Cartesian ones in the other) or numerical functions (a function can be tabulated, or defined by an equation).

Hardware

- Different microprocessors
- One more microprocessor in the control channel
- Two different types of computer

Project organization

- Different Software Design Teams
  (One in Paris and one in Toulouse as an extreme case)
- Two test sets designed by two different teams
- Different optimization goals: Timing performance vs program size

Inherent differences

- Hardware differences
- Some functions are required in only one channel
- Different input
- Different necessary precision

    12 bits for the control

    8 bits for the monitoring

- Function in the control channel, inverse function in the monitoring channel

Forced differences

- Different languages
    - PASCAL - ASSEMBLER
    - PLM - ASSEMBLER
    - Division of the instruction into two sub-sets
- Different automatic programming tools
- Different software specifications
- Different algorithms
- Different flowcharts
- A function can be tabulated or calculated
- Interrupts allowed in only one channel
- Trigonometric functions (polar coordinate vs Cartesian)

5

University Research

In this section two different research efforts are described. The first one deals with the most recent experiment with Recovery Blocks, the largest experiment and realization of this technique known so far.

The second paper describes the DEDIX-system, which can be used as an operating environment for N-version programs. This system, which was developed at UCLA in a multi year effort, proved to be portable. In a test installation, it is also running under Ultrix on Vax/785 at KfK Karlsruhe in Germany.

There have been many other experiments with software diversity of one kind or the other. Some of them are referenced in the following two papers. Besides those, the following shall be mentioned.

In an experiment by the University of Virginia and the University of California at Irvine 27 versions of a program - a part of a launch interceptor problem - were implemented by 27 teams. In the first analysis of 1 million test runs with these versions, the (hypothetical) assumption of complete independent failure behavior of the independently generated versions was shown to be false, as was generally expected [Knight 1985].

Further analysis of the data showed that N-version systems can reduce the probability of failure. The failure probability for a three version system decreased from 0.000 698 for a single version to 0.000 036 7 for a three version system [Knight 1986].

References

[Knight 1985]   J. C. Knight, N. G. Leveson, and L. D. St.Jean, "A Large Scale Experiment in N-Version Programming," in *Proc. 15th Intern. Symp. on Fault-Tolerant Computing FTCS' 15,* Ann Arbor, MI, USA: 19-21 June 1985, pp. 135-139.

[Knight 1986]   J. C. Knight and N. G. Leveson, "An Empirical Study of Failure Probabilities in Multi-Version Software," in *Proc. 16th Intern. Symp. on Fault-Tolerant Computing FTCS' 16,* Wien, A: 1-4 July 1986, pp. 165-170.

Tolerating Software Design Faults
in a Command and Control System

Tom Anderson
CSR, University of Newcastle upon Tyne

Peter A. Barrett
MARI, Newcastle upon Tyne

Dave N. Halliwell
CAP Scientific, London

Michael R. Moulding
Royal Military College of Science, Shrivenham

1. Introduction

The process of software development is usually described in terms of a progression from user requirements to the final code, passing through intermediate stages such as specification, design, and validation. Of course, progress through these stages is rarely unidirectional, and "final code" must be considered to be a misnomer given the demand for subsequent software maintenance. An engineering approach to software development should enable software to be produced on time, within budget, and in accordance with user requirements. One important aspect of these requirements concerns the **reliability** of the software. Software reliability requirements can be expressed in a number of ways, of which the simplest, perhaps, is to impose an upper limit on the measured rate of failure over a specified interval.

Given that reliability criteria can (and should) be imposed on software systems, how can these standards of reliability be achieved? Fortunately, there is a wide range of techniques available to the software developer, all intended to enhance software reliability. These techniques may be categorised as follows [Anderson 1981]:

1) techniques to avoid making mistakes - such as design methodologies and notations - referred to as **fault avoidance**;

2) techniques to find and remove mistakes - such as design reviews, code inspection, program analysis, testing, and verification, all followed by debugging or redesign - referred to as **fault removal**; and

3) techniques to cope with mistakes - defensive programming based on redundancy - referred to as **fault tolerance**.

The major obstacle impeding the construction of reliable software according to engineering principles is the shortage of data on the effectiveness of these various techniques. Fault tolerance techniques have played a major role in the development of reliable hardware systems [Carter 1985, Lala 1985], but have been much less widely used to cope with the possibly more serious problem of software reliability. Over the last ten years there has been considerable research activity addressing a range of issues in the field of software fault tolerance (see, for example, [Anderson 1981, Anderson 1987, Avižienis 1984, Slivinski 1984 and Welch 1983]). One outcome of this research has been the identification of specific notations and mechanisms for providing tolerance to software faults, including recovery blocks [Anderson 1976, Horning 1974, Lee 1978, Randell 1975] and N-version programming [Avižienis 1977, Chen 1978].

Nevertheless, the utilisation of this approach in practical systems remains rather limited, although dual-software systems have been constructed for a number of critical systems [Hagelin 1987, Martin 1982]. Again, the main reason for this may well be the lack of data on how effective this particular approach is in improving reliability. To date, the evaluation of software fault tolerance has either been performed by statistical modelling techniques [Bhargava 1981, Eckhardt 1985, Grnarov 1980, Laprie 1984, Migneault 1982, Scott 1983] or by empirical studies of multiple versions of software modules [Avižienis 1984, Kelly 1983, Knight 1986]. Both of these modes of evaluation have their limitations. The modelling approach is often bedevilled by unjustified assumptions and/or unquantifiable parameters, whereas the empirical approach has usually had to be applied to relatively small modules (because of cost considerations). Nevertheless, both approaches have indicated the potential for significant gains in software reliability from the use of fault tolerance techniques.

This chapter reports on a three-year project conducted at the University of Newcastle upon Tyne in conjunction with the Microelectronics Applications Research Institute (MARI). The aims of this project were:

1) to refine and develop software fault tolerance techniques for use in concurrent and real-time systems;

2) to confirm the utility of these techniques in a practical context;

3) to determine and quantify the effectiveness of the techniques for enhancing software reliability; and

4) to measure the cost and overheads incurred as a consequence of adopting fault tolerance.

In order that the results of the project could be considered applicable and relevant to current practical systems it was decided to implement, for evaluation purposes, an application system of reasonable scale, constructed by professional programmers to normal commercial standards. The application selected was a medium-scale naval command and control system, engineered to be as realistic as possible, but incorporating software fault tolerance capabilities based on recovery blocks and "conversations" [Horning 1974, Randell 1975].

An experimental programme was designed which involved executing the application software with a simulated tactical environment using a large number of action scenarios. Two modes of execution were available, depending on whether the fault tolerance features were enabled or disabled. Data from these experiments were analyzed to provide a number of quantitative assessments of the improvement in reliability arising from the use of fault tolerance. In fact, the results of this analysis suggest that software fault tolerance can prove very effective in coping with the consequences of faults in software.

This chapter provides an overview of the experimental configuration, describes the programme of experiments, summarises the data obtained from the experiments, and presents the analysis of and results derived from these data. Information on costs is briefly summarised in the conclusions. Project reports [Anderson 1985, Anderson 1984] provide full details of the experimental configuration and programme and also supply more details on costs.

2. Experimental System Configuration

The hardware configuration for the experimental system is illustrated in Fig. 1. Three DEC computers are employed to support the following sections of the system.

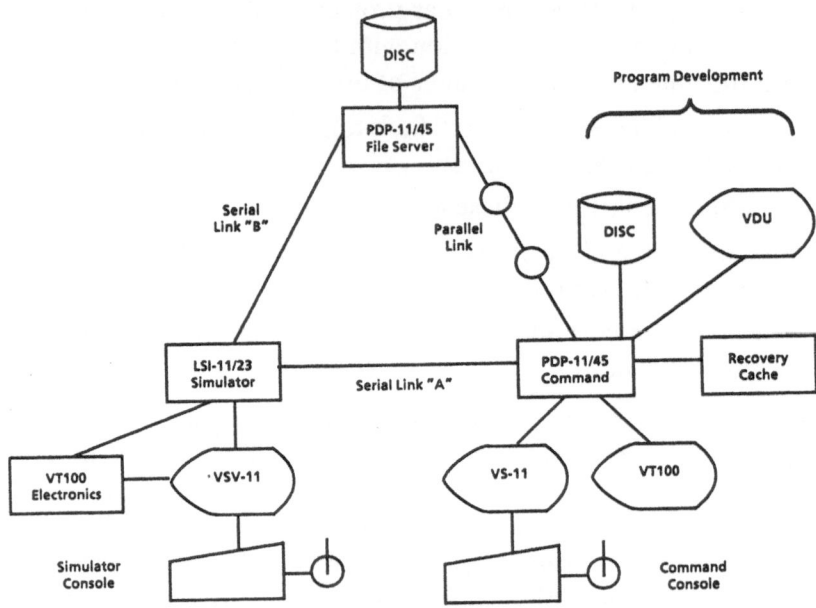

Fig. 1. Hardware configuration

A. Command (PDP-11/45)

This machine supports the command and control system itself. The software is written in the Coral language and runs under a project-developed MASCOT executive. (MASCOT is a design methodology for software construction and testing; the MASCOT executive is an operating system which supports and controls pseudoconcurrent processes and their interactions [Mascot 1980].) The command and control system was designed according to MASCOT techniques, and documented to the standard defined by JSP 188 (the U.K. Ministry of Defence standard for military systems); the involvement of the Royal Navy was sought to ensure that the system would be realistic in scale and functionality. The command and control system takes its input from simulated radar, sonar, and inertial navigation systems, displays the information from these sensors on a label plan display (LPD - a simulated radar display overlaid with track markers), and interacts with an operator, allowing him to conduct a "vectac" - a vectored attack on a hostile submarine by means of a helicopter armed with a torpedo. The command and control system consists of approximately 8000 lines of Coral source code, structured into 14 concurrent activities as indicated in Fig. 2. The command and control system and its interfaces are summarised pictorially in Fig. 3.

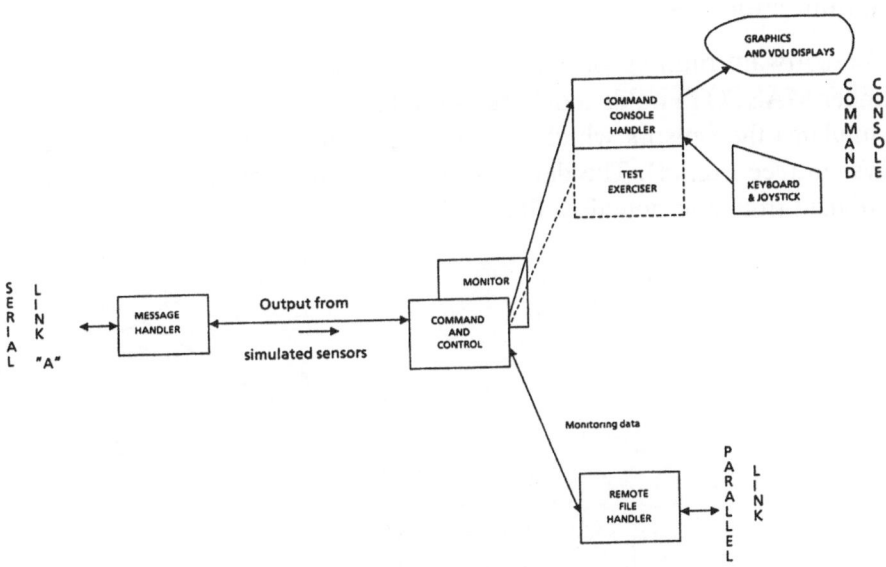

Fig. 2. Software Configuration for the PDP-11/45 (Command)

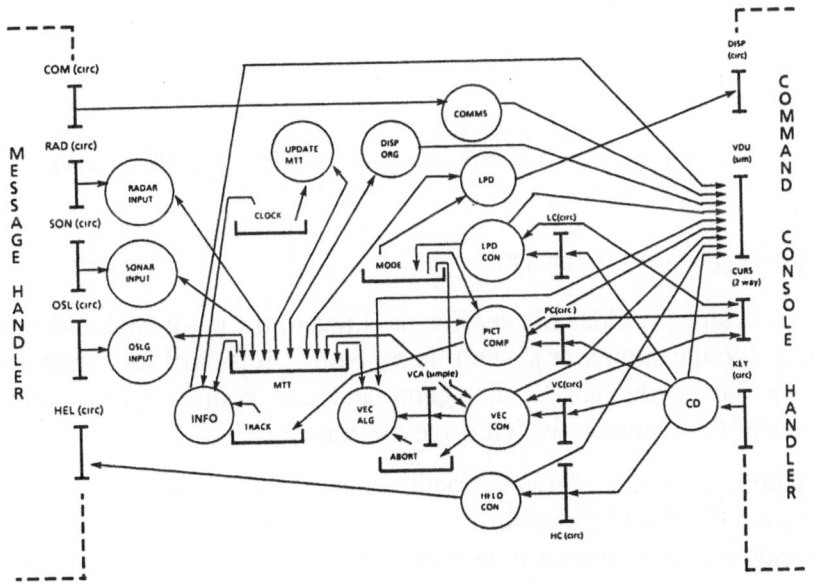

Fig. 3. MASCOT ACP Diagram of Command and Control Subsystem

B. Simulator (LSI-11/23)

The software running on this machine (again written in Coral and running under MASCOT) holds a data representation of the tactical environment and simulates the sensors which provide the input to the command and control system (see Fig. 4). The details of the tactical scenario to be used for a run are read from a remote data file.

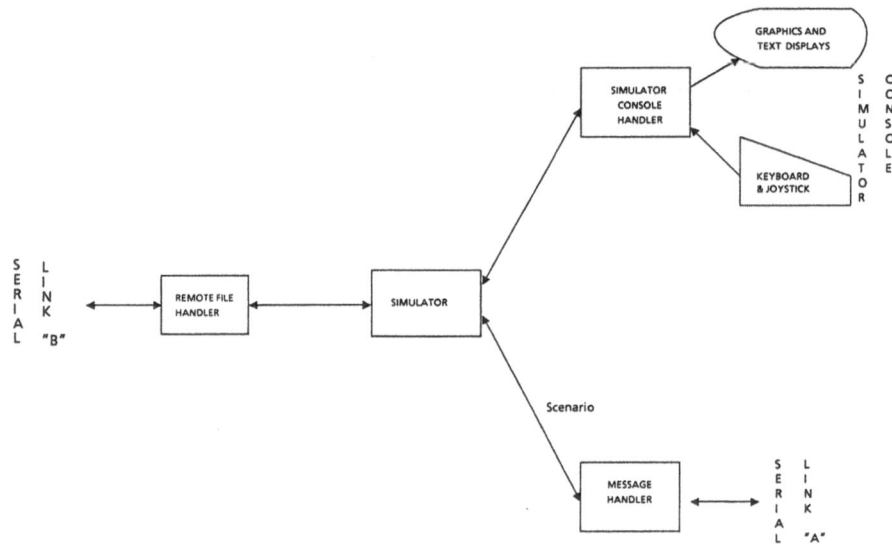

Fig. 4. Software Configuration for the LSI-11/23 (Simulator)

C. File Server (PDP-11/45)

This machine provides file service facilities to the command and control system and to the simulator system. In particular, it is used for logging the monitoring data generated by the command and control system and to hold details of the scenarios which drive the simulator.

In addition to the software mentioned above, various items of support software (MASCOT run-time executive, communications software, man-machine interface software, test and development software) were developed by the project.

Software fault tolerance was incorporated into the command and control software only, in the form of acceptance tests executed on completion of critical tasks, and as alternate modules of independent design which could be executed in the event of an acceptance test detecting a problem. Automatic state restoration was available to attempt to eradicate errors from the system. These facilities were extended to provide recoverable "dialogues" between multiple processes. A dialogue is an explicit embodiment of, and notation for, a restricted form of the concept of a conversation [Randell 1975], which is, in turn, an extension to concurrent systems of the recovery block technique. As such, the dialogue construct could be used to impose restraints, appropriate to a MASCOT system, on the provision of recovery to concurrent activities, while still permitting interprocess communication.

The MASCOT operating system was modified and extended to provide recovery capabilities for processes and for information recorded in the shared data areas used for process interaction. These recovery mechanisms utilised a special-purpose hardware device, called the recovery cache [Lee 1980], which enables state restoration to be performed very quickly (the recovery cache may be thought of as providing a highly optimised implementation of checkpointing for multiple processes).

The provision of design fault tolerance in the command and control software was constrained in the degree of diversity possible by the size of the implementation team. Three individuals made possible the "independent" design and construction of an acceptance test, a primary and a secondary alternate and this was the structure used in almost every case. Separate members of the team were not isolated (this would have been completely impractical in the context of the project) but the need for a strict discipline precluding cooperation or consultation on the design and implementation of the separate elements of the dialogue structures was very clearly understood. However, no attempt was made to enforce diversity other than by encouragement (and perhaps some general guidance) from the design consultant.

The actual tasks for which alternate algorithms and acceptance tests were provided were as follows:

1.   Associating an input radar message with an existing track.
2.   Updating a track using radar data.
3-6.   Tasks corresponding to the above, but for sonar and OSLG (own ship log and gyro) activities.
7.   Vectac mid-course control.
8.   Vectac course update calculation (nested in 7).
9.   Vectac final approach.
10.   Track classification.

11.   Track location.

12-13.   Track table house-keeping activities.

In addition to the above, an enclosing system level dialogue (utilising recovery and retry but without any alternate algorithm) formed an extra line of defence encompassing all of the above dialogues, thereby providing a degree of tolerance to faults detected by means of supplementary assertions embedded in code which was otherwise unprotected.

3. Experimental Programme

In order to measure the effectiveness of the software fault tolerance techniques in enhancing reliability, a series of experimental runs were performed using various tactical scenarios to drive the simulator system. Three phases of experimentation were conducted. For each phase of experimentation the application software was frozen; that is, no changes were made to the command and control software during a phase of the experiments.

The first phase of experiments involved two versions of the command and control system. In version one, the software fault tolerance was enabled and operated normally, whereas in version two, fault tolerance was disabled by the simple expedient of forcing all run-time checks to return a positive (i.e. OK) response.

It was originally intended that each experiment would consist of a pair of runs, one conducted with fault tolerance enabled, the other with fault tolerance switched off. The intention was that the unrecoverable run would proceed along a similar path to the fault-tolerant run until an event occurred. The unrecoverable run would then provide data on the consequences of that event in a fault-intolerant system. A test exerciser subsystem (automatic operator) was constructed to run in conjunction with the command and control software in an attempt to provide a consistent operator reaction and thereby ensure repeatability. Experience soon showed, however, that the system did not provide the levels of repeatability required for such a method: two runs started in a similar manner were likely to follow quite different paths. The causes of this lack of repeatability are well understood and centre around the unpredictability of the external interfaces to the command and control system. In particular, the communications protocol used to interface with the simulator (which uses a combination of check-sums, time-outs, and retransmissions to ensure that no messages are lost or corrupted) is such that the ordering of the stream of messages from the simulated environment cannot be guaranteed to be the same for two runs conducted under similar conditions.

Because of these problems, fewer fault-intolerant runs were performed, and only in the first experimental phase. However, the fault-intolerant runs were used to provide information which enabled accurate predictions to be made of the effects of any event in the fault-intolerant version. Furthermore, measurement of the reliability of the two versions enabled a direct confirmation to be obtained of the improvement in reliability due to fault tolerance. For phases two and three it was felt that our knowledge of the system was adequate to dispense with this confirmation, so all runs in these phases were performed with fault tolerance enabled.

In the second phase of experiments the same command and control software was used as for the first. In part, the intention was to confirm the results of phase one. More importantly, however, the first phase identified numerous problems with the MASCOT recovery software, and these were corrected for phase two. Since the success of the fault tolerance techniques is utterly dependent on the recovery mechanisms, the results from phase two should more accurately reflect the benefits possible from fault tolerance in practice.

In the third phase of experiments, the command and control software was modified by replacing a number of modules with new versions written by inexperienced programmers. These versions were expected to contain a greater number and wider range of faults than the original modules. Furthermore, where original modules were retained, the sequencing of alternates in recovery blocks was reversed, so that the backup alternates were used as primary alternates (and vice versa). Any faults in the recovery mechanisms identified during phase two were rectified before phase three.

Two further phases of experimentation were envisaged, and one of these was attempted. The intention was to evaluate the effectiveness of the fault tolerance techniques at different levels of reliability, and thereby investigate whether their effectiveness diminished (or increased) at higher reliability levels. To this end, all faults identified in the application system during phase one were rectified to yield a more reliable version of the command and control software. Unfortunately, this system proved too reliable, in that failure data were generated much too slowly. This phase of experimentation was therefore terminated unsuccessfully.

Time and financial limitations precluded the last phase of experimentation, in which it was planned to utilise an unreliable version of the application system derived from incompletely tested modules, which had been archived during the development of the command and control software.

Each phase of experiments consisted of a number of runs of the command and control system for which tactical scenarios were enacted on the simulator. Each run was monitored by the support system and was carefully observed by an operator. Each time an event occurred (an event is either a system failure or the detection of an error in the state of the system) the entire system would halt, and the operator would first log the incident and then analyze the error and attempt to identify the fault which caused it. The run would then be continued to see if fault tolerance would enable a failure to be averted, or if the failure would nevertheless occur. The system itself also recorded data to monitor events; the categorisation of events presented in the next section is based on these two sources of information. A run was considered to have finished when the scenario was completed, or when a failure occurred which prevented the system from continuing.

The tactical scenarios used to drive the simulator were prepared manually, with the assistance and guidance of Royal Navy personnel. Each scenario must define many conditions, such as the state of wind and tide, and specify the movement and attributes of friendly and hostile units, including an enemy submarine. The set of scenarios were intended to provide an approximation to the usage profile of the command and control application when "in action" against the submarine. Naval staff were asked to validate scenarios to ensure that they were realistic and typical.

4. Experimental Programme Results

The results from the experimental programme are presented in two sections; the first section gives a summary of the events which occurred in the fault-tolerant runs (for each of the three phases), whereas the second section presents overall statistics for the two versions of the system in phase one.

A. Summary of the Fault-Tolerant Runs

The results in this section consist of a summary of the events which occurred during the fault-tolerant runs of the experimental programme where an event is defined to be either an observed failure or the detection by internal checks of a suspected error in the state of the system.

In order to analyze the data from each run it was necessary to determine whether or not each event would have resulted in failure had the system contained no fault tolerance features. Usually, the answers to such questions were obvious, but whenever there was any doubt surrounding the outcome of a particular event in a fault-intolerant system, the option was available to run the system in fault-intolerant mode and attempt to recreate the event in question. The effects of the event would then be directly observable. This was not found to be necessary in phases two and three of the experiments.

The following categories were used to group events according to their outcome.

Events which yield an improvement in reliability over the fault-intolerant system:

1) events in which an error was detected, the system state was restored, and an alternate was successfully executed, thus averting failure.

Events for which no change in reliability is produced in comparison to the fault-intolerant system:

2) events in which the system state was restored unnecessarily, but no failure resulted;

3) events in which the system state was restored successfully, but then the system failed (as it would have done in the absence of fault tolerance);

4) events in which an unsuccessful attempt to restore the system state led to a system failure (but the system would have failed anyway);

5) events in which no error was detected and the system failed (as it would have done in the absence of fault tolerance).

Events which result in a deterioration in reliability in comparison to the fault-intolerant system:

6) events in which defective fault tolerance (an unsuccessful, unnecessary state restoration) caused the system to fail.

Events for which the outcome is uncertain:

7) events in which the effect on the system was unclear.

Table I enumerates the event counts for each of these seven categories, in each of the three phases of experiments. Over the entire programme of 163 runs of the fault-tolerant system, 250 events were analysed, and classified as shown in the first section of the table. The second section of Table I provides a breakdown of the 98 events which occurred as the first event of a run of the system (indicating that in 65 of the runs no events were recorded - these runs were therefore fruitless in terms of yielding evidence of the recovery capability of the fault tolerance techniques). The distinction between all events and first events is made to factor out any effects which might arise due to considering events which occurred after the fault-intolerant system would have failed - and could therefore be considered inadmissible in comparing the two versions of the system.

Table I. Event Counts

	Phase 1	Phase 2	Phase 3
Total Number of Runs:	43	60	60
Summary of All Events			
1) Recovery averting failure	40	34	91
2) Unnecessary recovery	4	6	4
3) Recovery followed by failure	0	5	4
4) Defective recovery	13	18	17
5) Failure with no recovery	0	0	0
6) Failure caused by recovery	4	4	0
7) Outcome unclear	4	1	1
Total events:	65	68	117
Summary of First Events			
1) Recovery averting failure	7	20	24
2) Unnecessary recovery	4	3	4
3) Recovery followed by failure	0	1	0
4) Defective recovery	9	10	5
5) Failure with no recovery	0	0	0
6) Failure caused by recovery	3	3	0
7) Outcome unclear	4	0	1
Total first events:	27	37	34

The data from Table I is analysed later in this chapter to produce the results summarised in Table III. Events in category 2 are ignored in that analysis, so it should be noted here that although unnecessary recoveries did take place, they occurred infrequently. Careful modification of an operational system would be expected to minimise the incidence of spurious recovery.

An attempt was made to categorise the different errors which led to recovery and failure events, but no clear picture emerged. The most common form of error detection (perhaps not surprisingly in a real-time system) was a form of timeout enforced by the MASCOT system. In particular, this prevented infinite looping when invalid circular data structures were erroneously formed in the main track table. Other common error categories were invalid track associations (usually to a deleted track), queueing errors, errors which generated hardware traps and errors caused by a defective acceptance test.

B. Comparative Data for Fault-Tolerant and Fault-Intolerant Runs

Data in the previous section relate solely to runs of the fault-tolerant version of the system, and assessment of the impact of fault tolerance on system reliability depends upon the analysis and categorisation of the events which took place. In this section, data are presented which enable a direct comparison to be made between the overall reliability of the two versions of the system. This may seem to provide a superior approach to comparative evaluation, but the reader is cautioned that due to a variety of factors (discussed below) the implications to be drawn from these data must be stated less firmly than those based on the data of the preceding section. Because of this limitation, the data for this section were only collected during phase one.

Table II presents a summary of phase one of the experiments for both versions of the command and control system. It records the total number of experimental runs of each version, the total elapsed time during execution of these runs, the total number of failures which occurred, and the number of runs which were completed without a failure of the command and control system (i.e. either completion of the scenario or premature termination due to a failure elsewhere).

Table II. Comparative Data

	Fault Tolerant	Fault Intolerant
Total runs	43	17
Total run time	50 4213s	28 057s
Total failures	19	25
Failure-free runs	24	8

A number of points must be taken into consideration when analysing the data in Table II.

1) First the nature of a run should be considered. A run begins with a period of relative inactivity, during which little other than object and screen updating and object classification takes place, and during which very few events occur. This is followed by a phase during which the system supplies the operator with information enabling him to guide an armed helicopter to engage a target submarine, referred to as a "vectac" (vector and attack), which constitutes a period of intense activity during which events are much more likely to occur. After the vectac the system returns to relative inactivity until either a further vectac takes place or the run is stopped. During the experimental programme, in order to restrict runs to a manageable duration, and to ensure an adequate rate of occurrence of events, the periods of inactivity before and after a vectac were artificially curtailed. This was done by running the simulation in "fast run" mode until shortly before the vectac was due to commence, then ending the run shortly after the vectac had completed (assuming that the system continued to run until this point). This curtailment has the effect that, since the system is likely to suffer few, if any, failures during periods of relative inactivity, any reliability measurements based on timing figures (for example MTBF) will appear far worse than they otherwise would. Thus, such figures might give a less favourable impression for the proportion of events successfully recovered.

2) The lack of repeatability between runs (discussed earlier), the consequent lack of a one-to-one correspondence between fault-tolerant and fault-intolerant runs, and the divergence of the two systems when an event occurs, means that the implications of a direct comparison between the two systems are not as definitive as are the experimental results presented in the previous section.

3) To ensure that these data corresponded to all of the experimental runs in phase one, judgements were made concerning the four events in category 7 (outcome unclear). These events were classified as two failures (one in each of categories 4 and 6) and two spurious recoveries. (The experimenters were reasonably confident that this classification was accurate, but the assignment was less certain than the original categorisation.)

4) The figures are heavily weighted by one particular run in which 16 of the total of 25 fault-intolerant failures occurred. This run was in no way a "freak"; all the failures were explained by known faults. However, the frequency of occurrence of such runs will clearly affect the overall system reliability. Unfortunately, there is insufficient data from the fault-intolerant runs to deduce the probability that such a run will occur.

5. Analysis of Results

A number of different approaches can be adopted for estimating the increase in reliability which can be attributed to the provision of fault tolerance in the command and control system. Three approaches, characterising different aspects of reliability, are presented in the following sections. The first approach is based on estimating the "coverage" achieved by the fault tolerance techniques; that is, what proportion of potential failures are successfully averted thanks to software fault tolerance? The second approach provides a direct estimate of the mean time between failures for both versions of the system, while the third quantifies the proportion of missions successfully completed for the two versions of the system.

A. Coverage Analysis

The principal measure of the effectiveness of software fault tolerance was taken to be the "coverage" factor of these techniques; that is, the proportion of failures which would have occurred in a fault-intolerant system that are successfully averted by means of fault tolerance. To be more precise, for situations in which the fault-intolerant system would fail, **coverage** represents the probability that the fault-tolerant system will nevertheless continue to operate without failing. The required probability can be easily estimated from the data of Table I, and thus relies solely on event counts. The coverage factor is calculated as the ratio of the number of failures averted (event category 1) to the number of potential failures (event categories 1, 3, 4, and 5). Events in category 2 (spurious recovery) and category 7 (unclear events) are disregarded. Events in category 6 (failures introduced by fault tolerance) cannot be ignored, but are excluded from the initial calculation.

Thus, considering all events in phase one of the experiments, the coverage achieved by fault tolerance is estimated to be 40/53, which is approximately 0.75. This is the maximum likelihood estimate. A Bayesian analysis using the beta distribution indicates that the value estimated can be asserted to exceed 0.67 with 90 percent confidence. These figures should be abated to take into account the four failures caused by fault tolerance. The simplest approach regards these failures as "own goals" and subtracts them from the successes of category 1. An amended coverage estimate of 0.68 is then obtained.

Table III presents these coverage estimates for the three phases of experimentation. The estimates have been calculated for the two sets of data, namely, all-event data and first-event data.

Table III. Failure Coverage

	Phase 1	Phase 2	Phase 3
All Events			
Raw coverage	0.75	0.60	0.81
Bayesian 90 percent point	0.67	0.52	0.77
Abated coverage	0.68	0.53	0.81
First Events			
Raw coverage	0.44	0.65	0.83
Bayesian 90 percent point	0.29	0.53	0.74
Abated coverage	0.25	0.55	0.83

B. Failure Rate Analysis

Simple arithmetic applied to the data in Table II yields the following results:

Failure rate for the fault-tolerant system:	1.36/h
Failure rate for the fault-intolerant system:	3.21/h
Ratio (fault-tolerant/fault-intolerant):	0.42

Making the standard, although often unjustified, assumption that the mean time between failures (MTBF) can be calculated as the reciprocal of the failure rate yields the following alternative presentation of these results:

MTBF for fault-tolerant system:	0.74h
MTBF for fault-intolerant system:	0.31h
Ratio (fault-tolerant/fault-intolerant):	2.36

These results may be compared to those of the previous section by using the change in failure rate to provide an estimate for the coverage of failures by means of fault tolerance.

Failure coverage: $(3.21-1.36)/3.21 = 0.58$

This value of 0.58 should be compared to the estimate of 0.68 (abated coverage, all events, phase one) presented in the previous section. The agreement is reasonably close, and the measurements are mutually supportive. However, it should be remembered that the comparison between the fault-tolerant and fault-intolerant runs is by no means exact because of the inability to precisely repeat any individual run. In phases two and three, the failure rate for the fault-tolerant system improved to 0.88/h and 0.58/h (MTBF 1.14h and 1.72h).

C. Successful Missions

A further comparison between the two versions of the system may be made by examining the proportion of runs which were completed without a failure arising from the command and control system. Again, from Table II, it can be seen that the fault-tolerant system is more successful, although the improvement is much less marked.

Proportion of fault-tolerant runs which
completed without failing: 56%

Proportion of fault-intolerant runs which
completed without failing: 47%

Ratio (fault-tolerant/fault-intolerant)   1.19

6. Discussion and Conclusion

The results of the previous section show clearly that for this application, in these experiments, the incorporation of software fault tolerance has yielded a substantial increase in reliability. Over the entire programme of experiments (phases 1-3), the event counts of Table I show that 222 failures could have occurred due to "bugs" in the software of the command and control system. But of these 222 potential failures, only 57 (9 in category 3, 48 in category 4) actually happened - the other 165 were masked by the use of software fault tolerance (category 1). This represents an overall success rate of 74 percent. (The same calculation restricted to first events yields the slightly lower figure of 67 percent.)

Examination of the results from the first phase of experiments suggested that much better results could be achieved if the underlying recovery mechanisms could be brought to an adequate standard of reliability. Essentially, the project was relying on prototype recovery mechanisms (the recovery cache and the MASCOT recovery software) to support the provision of fault tolerance at the application level. This situation would most certainly not be typical of an operational system, where the recovery facilities should be at least as reliable as the hardware itself. It was hoped that improvement to the recovery routines for phase two would produce improved results, but in fact this effect was not observed until phase three. Projections suggest that with further improvements to the recovery software a coverage factor of over 90 percent could have been achieved.

The discrepancy between the results for all events and first events is very marked for phase one, but is minimal in phases two and three. The most likely explanation is that the results for all events in phase one are rather better than they would otherwise be as a result of multiple recovery successes occurring in sequence. This phenomenon did occur in one spectacular case in phase one where a series of 12 successful recoveries in rapid succession in a single run helped boost the figures (and, to some extent, project morale).

Of course, these gains were achieved at a cost, paid in capital costs to support fault tolerance, development costs to incorporate fault tolerance, and run-time and storage costs to make use of fault tolerance.

The capital cost for supporting fault tolerance consisted of the costs of acquiring a hardware recovery device, for developing recovery software and incorporating this in the MASCOT operating system, and devising an interface by which dialogues and recovery blocks could interact with the operating system. The project expended approximately 1000 man-hours on these tasks, but the aim for the future would be that recovery facilities should be available on systems for critical applications on payment of a limited premium to the system manufacturer.

The supplementary development cost of incorporating fault tolerance in the command and control system was approximately 60 percent. This covered the provision of the acceptance tests and alternate modules used in recovery blocks and dialogues. The figure of 60 percent is probably rather high, reflecting the novelty of the techniques employed and their unoptimised utilisation in this particular application. Against the increased development cost must be offset any gains resulting from economies in testing the software.

Overheads in system operation were measured as: 33 percent extra code memory, 35 percent extra data memory, and 40 percent additional run-time (although the system still had to meet its real-time constraints). The run-time overhead was incurred largely as a penalty for the synchronisation of processes for consistent recovery capability: data collection for state restoration purposes only contributed about 10 percent of the run-time overhead. By tuning the system to optimise real-time response this overhead could be substantially reduced.

Our overall conclusion is that these experiments have shown that by means of software fault tolerance a significant and worthwhile improvement in reliability can be achieved at acceptable cost. We look forward to an independent confirmation of this result, preferably in the context of a system to be used in earnest.

Acknowledgement

This chapter is a revised version of a paper originally published in IEEE Transactions on Software Engineering, Vol. SE-11, No. 12, Dec. 1985, pp. 1502-1510. Permission to include it in this volume is hereby acknowledged.

References

[Anderson 1985] T. Anderson and P. A. Barrett, "Fault Tolerance Project Report: Results and Conclusions from the Second and Third Experimental Programmes", University of Newcastle upon Tyne, Proj. Rep. 4844/DD.17/3, 1985.

[Anderson 1976] T. Anderson and R. Kerr, "Recovery Blocks in Action: A System Supporting High Reliability", in Proc. Second Int. Conf. Software Eng., San Francisco, CA, 1976, pp. 447-457.

[Anderson 1981] T. Anderson and P. A. Lee, Fault Tolerance: Principles and Practice. Englewood Cliffs, NJ: Prentice-Hall 1981.

[Anderson 1984] T. Anderson et al., "Fault Tolerance Project Report: Results and Conclusions from the Experimental Programme", University of Newcastle upon Tyne, Proj. Rep. 4844/DD.17/2, 1984.

[Anderson 1987] T. Anderson, "Design Fault Tolerance in Practical Systems", in Software Reliability: Achievement and Assessment, B. Littlewood (Ed.), Oxford: Blackwell Scientific, 1987.

[Avižienis 1977] A. Avižienis and L. Chen, "On the Implementation of N-Version Programming for Software Fault Tolerance During Program Execution", in Proc. COMPSAC 77, Chicago, IL, 1977, pp. 149-155.

[Avižienis 1984] A. Avižienis and J. P. J. Kelly, "Fault Tolerance by Design Diversity: Concepts and Experiments", Computer, vol. 17, pp. 67-80, August 1984.

[Bhargava 1981] B. Bhargava, "Software Reliability in Real-Time Systems", in Proc. NCC, Chicago, IL, 1981, pp. 297-309.

[Carter 1985] W. C. Carter, "Hardware Fault Tolerance", in Resilient Computing Systems, T. Anderson (Ed.), New York: Wiley, 1985, pp. 11-63.

[Chen 1978] L. Chen and A. Avižienis, "N-Version Programming: A Fault-Tolerance Approach to Reliability of Software Operation", in Dig. FTCS-8, Toulouse, France, 1978, pp. 3-9.

[Eckhardt 1985] D. E. Eckhardt and L. D. Lee, "A Theoretical Basis for the Analysis of Multi Version Software Subject to Coincident Errors", IEEE Trans. Software Eng., vol.

SE-11, pp. 1511-1517, December 1985.

[Grnarov 1980] A. Grnarov et al., "On the Performance of Software Fault Tolerance Strategies", in Dig. FTCS-10, Kyoto, Japan, 1980, pp.251-253.

[Hagelin 1987] G. Hagelin, "ERICSSON Safety Systems for Railway Control", in this volume.

[Horning 1974] J. J. Horning et al., "A Program Structure for Error Detection and Recovery", in Lecture Notes in Computer Science 16. New York: Springer-Verlag, 1974, pp. 171-187.

[Kelly 1983] J. P. J. Kelly and A. Avižienis, "A Specification Oriented Multiversion Software Experiment", in Dig. FTCS-13, Milan, Italy, 1983, pp.120-126.

[Knight 1986] J. C. Knight and N. G. Leveson, "An Empirical Study of Failure Probabilities in Multi-Version Software", in Dig. FTCS-16, Vienna, Austria, 1986, pp. 165-170.

[Lala 1985] P. K. Lala, Fault Tolerant and Fault Testable Hardware Design. Englewood Cliffs, NJ: Prentice Hall, 1985.

[Laprie 1984] J.-C. Laprie, "Dependability Evaluation of Software Systems in Operation," IEEE Trans. Software Eng., vol. SE-10, pp. 701-714, June 1984.

[Lee 1978] P. A. Lee, "A Reconsideration of the Recovery Block Scheme", Comput. J., vol. 21, no. 4, pp. 306-310, 1978.

[Lee 1980] P. A. Lee et al., "A Recovery Cache for the PDP-11", IEEE Trans. Comput., vol. C-29, pp. 546-549, June 1980.

[Martin 1982] D. J. Martin, "Dissimilar Software in High Integrity Applications in Flight Controls", in Proc. AGARD Symp. Software Avionics, The Hague, The Netherlands, 1982, pp. 36:1-36:13.

[Mascot 1980] Mascot Suppliers Ass., The Official Handbook of MASCOT, Royal Signals and Radar Establishment, Malvern, England, 1980.

[Migneault 1982] G. E. Migneault, "The Cost of Software Fault Tolerance", in Proc. AGARD Symp. Software Avionics, The Hague, The Netherlands, 1982, pp. 37:1-37:8.

[Randell 1975] B. Randell, "System Structure for Software Fault Tolerance", IEEE Trans. Software Eng., vol. SE-1, pp. 220-232, June 1975.

[Scott 1983] R. K. Scott et al., "Modelling Fault-Tolerant Software Reliability", in Proc. 3rd Symp. Reliability Distrib. Software Database Syst., Clearwater Beach, FL, 1983, pp. 15-27.

[Slivinski 1984] T. Slivinski et al., "Study of Fault Tolerant Software Technology", Mandex Inc., Rep. NASA Langley Res. Cen., 1984.

[Welch 1983] H. O. Welch, "Distributed Recovery Block Performance in a Real-Time Control Loop", in Proc. Real-Time Sys. Symp., Arlington, VA, 1983, pp. 268-276.

DEDIX 87 - A Supervisory System for Design Diversity Experiments at UCLA

Algirdas Avižienis, Michael R. T. Lyu, Werner Schütz,
Kam-Sing Tso, Udo Voges

Dependable Computing and Fault-Tolerant Systems Laboratory
UCLA Computer Science Department
University of California
Los Angeles, CA 90024, USA

Abstract

To establish a long-term research facility for further experimental investigations of design diversity as a means of achieving fault-tolerant systems, the DEDIX (DEsign DIversity eXperiment) system, a distributed supervisor and testbed for multi-version software, was designed and implemented by researchers at the UCLA Dependable Computing and Fault-Tolerant Systems Laboratory. DEDIX is available on the Olympus local network, which utilizes the Locus distributed operating system to operate a set of several VAX 11/750 computers at the UCLA Center for Experimental Computer Science. DEDIX is portable to any machine which runs a Unix operating system. The DEDIX system is described and its applications are discussed in this paper. A review of current research is also presented.

1. Introduction

By early 1970s significant progress had been made in the development of systems that tolerate physical faults that are due to random failures of components or physical interference with the hardware of a system. At that time it became clear that design faults, especially as represented by software "bugs", presented the next challenge to the researchers in fault tolerance. A research effort to attain tolerance of design faults by means of multi-version software was started at UCLA in early 1975. The method was first described as "redundant programming" at the April 1975 International Conference on Reliable Software in Los Angeles [Avižienis 1975], and was renamed as "N-version programming" in the course of the next two years [Avižienis 1977]. The name "Multi-Version Software" (MVS) has also been used. The entire UCLA design diversity research effort through mid-1985 has been summarized in [Avižienis 1985b].

The N-version programming approach to fault tolerant software systems employs functionally equivalent, yet independently developed software components. These components are executed concurrently under a supervisory system that uses a decision algorithm based on consensus to determine final output values. From its beginning in 1975, the fundamental conjecture of the MVS approach at UCLA has been that errors due to residual software faults are very likely to be masked by the correct results produced by the other versions in the system. This conjecture does not assume independence of errors, but rather a low probability of their concurrence and similarity. MVS systems achieve reliability improvements through the use of redundancy and diversity. A "dimension of diversity" is one of the independent variables in the development process of an MVS system. Diversity may be achieved along various dimensions, e.g., specification languages, specification writers, programming languages, programmers, algorithms, data structures, development environments, and testing methods.

The scarcity of previous results and an absence of formal theories on N-version programming in 1975 led to the choice of an experimental approach: to choose some conveniently accessible programming problems, to assess the applicability of N-version programming, and then to proceed to generate a set of programs. Once generated, the programs were executed as N-version software units in a simulated multiple-hardware system, and the resulting observations were applied to refine the methodology and to build up the concepts on N-version programming. The first detailed assessment of the research approach and a discussion of two sets of experimental results, using 27 and 16 independently written programs from a software engineering class, was published in 1978 [Chen 1978].

This exploratory research demonstrated the practicality of experimental investigation and confirmed the need for high quality software specifications. As a consequence, the first aim of the next phase of UCLA research (1979-82) was the investigation of the relative applicability of three distinct software specification techniques: formal (OBJ), semiformal (PDL), and in

English.

Other aims were to investigate the types and causes of software design faults, to propose improvements to software specification techniques and their use, and to propose future experiments for the investigation of design fault tolerance in software and in hardware [Kelly 1983, Avižienis 1984].

In the course of the experiments at UCLA it became evident that the usual general-purpose campus computing services were poorly suited to support the systematic execution, instrumentation, and observation of N-version fault-tolerant software. In order to provide a long-term research facility for experimental investigations of design diversity as a means of achieving fault-tolerant systems, researchers at the UCLA Dependable Computing and Fault-Tolerant Systems Laboratory have designed and implemented the prototype DEDIX (DEsign DIversity eXperiment) system, a distributed supervisor and testbed for multiple-version software [Avižienis 1985a]. DEDIX is supported by the Olympus Net local network at the UCLA Center for Experimental Computer Science which utilizes the UNIX-based Locus distributed operating system to operate a set of VAX 11/750 computers.

The purpose of DEDIX is to supervise and to observe the execution of N diverse versions of an application program functioning as a fault-tolerant N-version software unit. DEDIX also provides a transparent interface to the users, versions, and the input/output system, so that they need not be aware of the existence of multiple versions and recovery algorithms. The prototype DEDIX system has been operational since early 1985. Several modifications have been introduced since then, most of them intended to improve the speed of the execution of N-version software. The first major test of DEDIX has been the experimentation with the set of five programs produced at UCLA for the NASA-sponsored Four-University N-version software project. A complete overview of the structure and of the applications of DEDIX at UCLA is presented in this paper.

1.1 Functional Requirements of DEDIX

The principal functional requirements of DEDIX are as follows:

Distribution: the versions must be able to execute on separate physical sites in order to take advantage of physical isolation between sites, to benefit from parallel execution, and to survive a crash of a minority of sites.

Transparency: the application programmer must not be required to write special software to take care of the multiplicity, and a version must be able to run in a system with any allowed value of N without modifications.

Decision making: a reliable *decision algorithm* that determines a single *consensus result* from the multiple version results must be provided. The algorithm must be able to deal with specified allowable differences in numerical values and with slightly different formats (e.g. misspellings) in human-readable results; additionally, the user must be able to choose between different decision algorithms and even -- with some more effort -- be able to incorporate a special decision algorithm of his own. If a consensus cannot be obtained, an alternate decision must be provided.

Recovery and/or reconfiguration: DEDIX must support recovery attempts for the minority of disagreeing versions. It also must implement reconfiguration decisions that remove failed versions or sites when recovery is not available or does not succeed. When a consensus does not result, an alternate outcome (safe shutdown, or invocation of a backup system) must be implemented.

Environment: DEDIX must run on the distributed Locus environment at UCLA [Popek 1981] and must be easily portable to other UNIX systems. DEDIX must be able to run concurrently with all other normal activities of the local network.

1.2 Related Research

The DEDIX system can be considered as a network-based generalization of SIFT [Wensley 1978] that is able to tolerate both physical and design faults in software and in hardware. Both have similar partitioning, with a local executive and a decision algorithm at each site that processes broadcast results, and a copy of the global executive at each site that takes consistent recovery and reconfiguration decisions by majority vote. DEDIX is extended to allow some diversity in results and in version execution times. SIFT is a frame synchronous system that uses periodically synchronized clocks to predict when results should be available for a decision. This technique does not allow the diversity in execution times and unpredictable delays in communication that can be found in a distributed N-version environment, especially when it is shared with other jobs. Instead, a synchronization protocol is used in DEDIX which does not make reference to global time within the system.

Another approach to fault-tolerant software is the *Recovery Block* technique, in which alternate software versions are organized in a manner similar to the dynamic redundancy (standby sparing) technique used in hardware [Anderson 1981]. The objective of the recovery block technique is to perform software design fault detection during runtime by an acceptance test performed on the results of one version, as opposed to comparing results from several versions. If the test fails, an alternate version is executed to implement recovery. Several major research activities related to N-version programming and recovery block techniques have been reported [Anderson 1985, Cristian 1982, Kelly 1986, Kim 1984, Ramamoorthy 1981, Voges 1982].

2. Functional Description of the DEDIX System

2.1 Services and Structure

DEDIX together with the diverse program versions has the ability to tolerate software design and implementation faults. DEDIX and the versions interact with each other and with their environment, i.e., a user, so that together they can be seen as a fault-tolerant *multi-version system*. DEDIX itself is a *supervisor* that does not add any application relevant functions to the system.

The purpose of DEDIX is to supervise and to observe the execution of N diverse versions of an application program functioning as a fault-tolerant N-version software unit. DEDIX also provides a transparent interface to the users, the versions, and the input/output system so that they need not be aware of the existence of multiple versions and recovery algorithms. An abstract view of DEDIX as a multiversion system with N versions is given in Fig. 1. Generally speaking, DEDIX provides the following services:

- it handles communications from the user and distributes them to all active versions;
- it handles requests from the versions to have their results (cc-vectors) processed, and returns consensus results to the versions and to the user;
- it executes decision algorithms and determines consensus results, or invokes alternate decisions if a consensus does not exist;
- it manages the input and output operations for the versions; and
- it makes reconfiguration and recovery decisions about the handling of faulty versions.

Partitioning of DEDIX. The DEDIX system can be located either in a single computer that executes all versions sequentially, or in a multicomputer system running one (or more) versions at each site. If DEDIX is supported by a single computer, it is vulnerable to hardware and software faults that

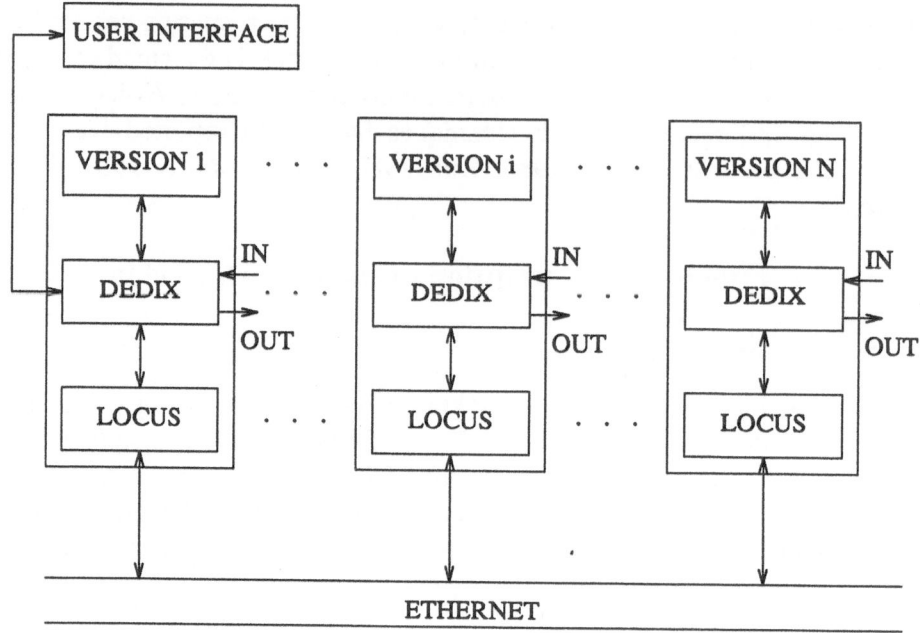

Fig. 1. DEDIX as a Multiversion System

affect the host computer, and the execution of *N*-version software units is relatively slow. In a computer network environment, the system is partitioned to protect against most hardware faults. This has been done by providing each site with its own local DEDIX software, while an intersite communication facility is common to all computers. The DEDIX design is suitable for any specified number $N \geq 2$ of sites and versions, and currently accommodates the range $2 \leq N \leq 20$, with provisions to reduce the number of sites and to adjust the decision algorithm upon the failure of a version or a site.

The manifestation of faults. A hardware or software fault will affect a program version and it may also affect the underlying system. DEDIX is designed to be able to identify a malfunctioning site and to tolerate both cases of fault effects, provided that the errors can be detected. In the first case, when the errors and the faults can be isolated to a version only, the site will attempt to recover the internal state of the local version with decision results. In the second fault case, the site usually will not be able to recover by itself and a global reconfiguration decision is necessary. All *version faults* will manifest themselves as either "incorrect results", or "missing results".

For example, a missing result from a site can be caused by an erroneous version, which is in an infinite loop, a deadlocked operating system, a hardware fault causing an error in the DEDIX software, etc. A missing result at a site might also be caused by an excessive communication delay, i.e., the result is produced but does not reach the other sites in time. In this case, the sending site will detect the discrepancy between what it sent and what the other sites observed.

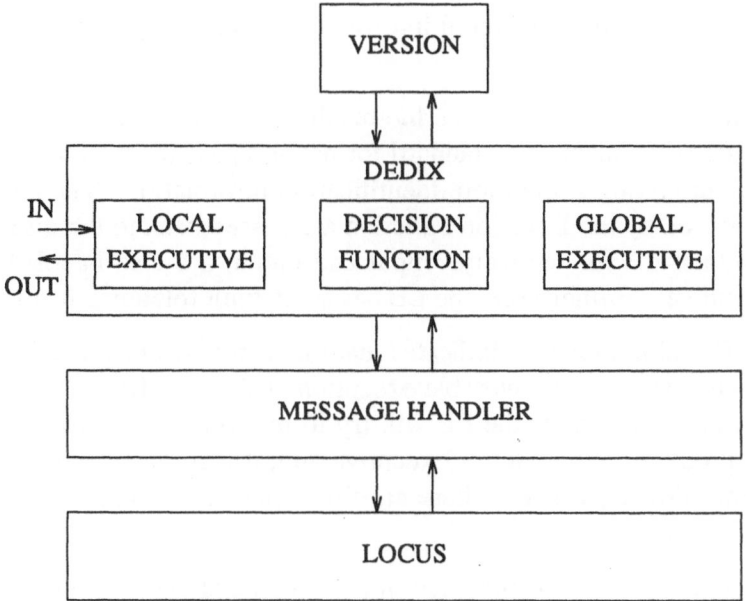

Fig. 2. Functional Structure of DEDIX

Structure. The services of DEDIX are partitioned into three functional modules as shown in Fig. 2 and described in more detail next. They are:

- a *Local Executive* (LE), which is the DEDIX interface to the local version and provides input/output facilities. It also supervises local recovery or reconfiguration actions for the local version.

- a *Decision Function* (DF), which compares a set of results from the different versions to produce a decision result, which is either a consensus result or a decision that a consensus does not exist.

- a *Global Executive* (GE), which monitors the execution at all sites and versions and supervises global recovery and reconfiguration in case of version failures or a lack of consensus.

2.2 The Local Executive

The Local Executive (LE) contains the DEDIX interface to the version. The version interacts with DEDIX via calls to cross-check functions (cc-functions) and recovery points [Tso 1987b]. The incorporation of these calls is the main adjustment a user has to make in comparison to running his program in a normal (single version) environment. The exact form of these calls is described in Section 3.2. The point of interaction is called cross-check point (cc-point) and the transfered information accordingly cross-check vector (cc-vector).

At cc-points, the cc-functions take the results from the version in form of a cc-vector, translate them to a standard format and pass them to the Decision Function after adding some more identification information. The consensus results produced by the Decision Function are passed back to the disagreeing versions by the LE for recovery. Input and output are also handled by the Local Executive. Furthermore, the LE has some fault tolerance features.

When the Decision Function indicates that the consensus result is not unanimous, or when some unrecoverable exception is signaled from the local version or some other source, the LE will try to recover locally from the fault, report the problem to the Global Executive and, if it is considered as fatal to the site, shut down the site. There are three classes of exceptions that are considered, as discussed below.

Functional exceptions are specified in the functional description of DEDIX and are independent of the implementation. Among them are deviations from an unanimous result, the case when a communication link is disconnected, and the case when a cc-vector is completely missing. For these exceptions the Local Executive will attempt to keep the site active, possibly terminating the local version, while keeping the input/output operating.

Implementation exceptions are dependent on the specific computer system, language, and implementation technique chosen. All UNIX signals, such as segmentation faults, process termination, invalid system call, etc., belong to this class. Other examples are all the exceptions defined in DEDIX, such as signaling when a function is called with an invalid parameter, or when an inconsistent state exists. Most of these exceptions will force an orderly shutdown of a site in order to be able to provide data for analysis.

Exceptions generated by the local version. The local version program is likely to include facilities for exception handling, and some of the exceptions may not be recoverable within the version. These exceptions are sent to the Local Executive which will terminate the local version, while keeping the site alive.

2.3 The Decision Function

The Decision Function is used to determine a single consensus result from the N version results. The Decision Function may utilize only a subset of all N results for a decision; for example, the first result that passes an acceptance test may be chosen. In case a consensus result cannot be determined, a higher level recovery procedure needs to be invoked, that is determined by the Global Executive.

DEDIX provides a generic decision algorithm which may be replaced by the user's custom algorithm, provided that the interfaces are preserved. This allows application-specific decision algorithms to be incorporated in those cases where the standard decisions are inappropriate, or insufficiently precise.

The current decision algorithm searches for a consensus by applying one of the following comparisons to the version results:

(1)   **exact (bit by bit)** - allowing an identical match only;

(2)   **numerical** - integer and real number comparisons with an allowed range of deviation for "similar" results;

(3)   with **"cosmetic"** corrections - allowing for minor (defined) character string differences that may be caused by misspelling or character substitution.

2.4 The Global Executive

The Global Executive (GE) is activated when a recovery point (r-point) is executed. Each r-point has a unique r-point identifier (rp-id) [Tso 1987b]. At first, the GE performs the following actions to determine if global recovery is necessary: 1) compares the rp-ids delivered by the versions, 2) exchanges error reports with other Global Executives, and 3) determines which versions have failed, i.e., disagree with the consensus.

Error reports. Every Global Executive has an error report table, with one entry for each site. This entry is an error counter for that site. The GE increments the counter for a site, whenever that site has either a disagreeing or missing result at a cc-point. This means that the GE does distinguish between a missing result and a delayed result. Since sites might get different numbers of results due to varying communication delays, sites may have somewhat different error reports. The exchange and comparison of error reports ensure a consensus among the GEs at various sites on which versions have failed. If no failed version is detected, the GE merely resets the error report table and the versions continue their execution. Otherwise, global error recovery is initiated.

Two types of failed versions are distinguished: 1) those that have detected errors at the cc-points, and 2) those with incorrect or missing rp-ids. Each Global Executive of a good version signals the *state-output exception handler* of its local version to output the internal state at that rp-id. These states are compared by the Decision Function to obtain the Decision State. Each failed version of the first type is recovered by invoking its *state-input exception handler* to input the Decision State. After the exchange of internal states, actions of the global error recovery are completed and execution of the versions is resumed. A failed version of the second type is first aborted by its Global Executive. The version is then restarted by its GE at the r-point with the decision rp-id. The restarted version also inputs the Decision State by invoking the state-input exception handler before its execution is resumed.

The reconfiguration decision. If a version has produced errors at two or more consecutive r-points, reconfiguration (by shutdown) needs to be initiated. If the shutdown applies to a site, each Local Executive instructs its message receivers to stop receiving from that site. If the shutdown applies to a version, its Local Executive terminates the local version and stops sending messages. In both cases, the new number of expected results is adjusted accordingly by the Decision Function at all sites. After a version is shut down, the site will still collect messages and operate input/output, but it will not deliver them to the local version. The Decision Function and the Global Executive at a site are not affected if only the local version is shut down.

2.5 The Message Handler

Since Locus does not supply all the message handling routines needed for DEDIX, an interface between the described three functional modules of DEDIX and the Locus operating system is necessary. This message handler (MH) interface consists of two layers: the *Synchronization layer* and the message *Transport layer*, where the Synchronization layer is the DEDIX-dependent part of the message handler, while the message Transport layer is DEDIX-independent and depends on the service provided by the underlying operating system.

2.5.1 The Synchronization Layer

For each physically distributed site, the Synchronization Layer (SL) broadcasts results (using the Transport Layer) and collects messages with the version results ("cc-vectors") from all other sites. The SL only accepts results that are both broadcast within a certain time interval and that arrive within the same time interval. The collected results are delivered to the functional modules at the site. The SL accepts a new set of version results when every site has confirmed that all or enough of the previous results have been delivered.

The sites of DEDIX need to be event-synchronized in order to ensure that results from corresponding cc-points are compared. Otherwise, if two sets of results from two different cc-points are compared, the Decision Function might wrongly conclude that some of the versions or sites are faulty. Traditionally, this synchronization has been obtained by referring to a common clock or set of clocks. The SIFT system [Melliar-Smith 1982] is one example of such a clock synchronous system. In SIFT it is predicted *when* the results should be available for a comparison. To ensure that the results are available in SIFT, several design measures are taken to eliminate all unpredictable delays, such as using a fully connected communication structure, using strict periodic scheduling, not allowing external interrupts (only clock interrupts are allowed for scheduling), and regularly synchronizing the clocks.

The local network system and the diverse versions have the following characteristics which make the clock synchronous technique impractical in DEDIX:

- the versions can have different *execution times* between the cross-check points;

- the versions *will run concurrently* with other network activities, which means that sites temporarily can be heavily loaded, and hence prolong the time to execute some versions;

- the Ethernet communication network has inherently varying message transport delays.

A synchronization protocol is designed to provide the event-synchronization service. It ensures that the results that are compared by the Decision Function are from the same cross-check point (cc-point) in each version. The versions are halted until all of them have reached the same cc-point, and they are not started again until the results are exchanged and a decision is made. To be able to detect versions that are in an infinite loop, or otherwise too slow, a time-out technique is used by the protocol.

The use of this synchronization protocol is based on the assumptions that:

(a)   correctly working versions produce exactly the same number of cc-vectors in the same order;

(b)   correctly working versions have compatible execution times, i.e., they will produce results within a specified time-out interval;

(c)   "missing" or disagreeing results do not exist for a majority of versions.

The specification and verification of the protocol is described in [Gunning-berg 1985].

Time-out function. The only way to detect that a version did not produce a result when it was supposed to, or that the result is "stuck" somewhere in the communication system, is to use a time-out function, i.e., to require that every version must produce a result within a time-out interval. Two time-out techniques have been considered. The first technique is similar to the time-acceptance test in the recovery block technique. A time-out function is started at the beginning of each segment of computation and all versions must produce results within the specified time interval in order to pass the time-acceptance test. The length of the interval can either be adjusted to each segment of computation or to a "worst case" interval for all segments.

In the second technique, the time-out interval is started when a majority of results have arrived at a site. For example, the time-out is started when the third result arrives in a configuration with five active versions. This technique is based on a comparison between relative execution times instead of using an absolute time, as in the first technique. The time-out is of course terminated if all results arrive before the time-out interval expires. A malfunctioning version sending results too early will not cause any problems, since they will not start the time-out. Interestingly, the problem is similar to "comparing results with skew": the median number (result number 3 out of 5) constitutes the closest to the "ideal value" and the skew corresponds to the time interval. One advantage with this technique, compared to the previous, is that there is no need to assign an individual time-out for each segment of computation. This is an advantage, since the execution time might depend on an a priori unpredictable input, which might put the computation into a loop of long duration. Furthermore, since the time between different cc-points may vary and the sequence of the cc-points is not predetermined, the synchronization would need complex information to adjust the individual time-out intervals.

Both techniques can exist together in DEDIX, and the choice may depend on the application, the input/output, the computing environment, and the real time requirements. Both techniques require that version computations should start almost at the same time at each site and that user input also must arrive within the defined time interval. In the current DEDIX system, the second technique is implemented, due to the operating environment and the type of computations. The time interval is set by the user and can be quite wide, since all versions are suspended until the time interval has expired or until all results have arrived. This suspension is possible since currently there is no real time requirement within DEDIX. The system would need some modifications to accommodate the time-acceptance test technique.

2.5.2 The Transport Layer

The Transport Layer (TL) controls the communication of messages. It hides the system primitives that are actually employed from its user modules. The TL makes sure that no message is lost, duplicated, damaged, or misaddressed, and it preserves the ordering of sent messages.

The requirements of the Transport Layer are specified in terms of response time, throughput, and reliability of service. In order to satisfy the reliability requirement, in most practical situations a redundant communication structure needs to be used. Currently, a single ring structure of inter-process UNIX pipes is employed due to the limitation on the number of pipes per process. Since this implementation does not tolerate a site crash, a redundant interconnection structure is under implementation at the present time.

2.6 Possible Configurations of DEDIX

Given the fundamental functional modules of DEDIX that were described above (LE: Local Executive; DF: Decision Function; GE: Global Executive; SL: Synchronization Layer; TL: message Transport Layer), several configurations of DEDIX can be implemented.

    a)   For standard operation all three functions, LE, DF, and GE, reside on the same site, and all are present in the same number as the versions. This leads to the standard DEDIX configuration as shown in Fig. 3, with one SL and one TL servicing all three functions.

    b)   It is possible to have fewer GE modules than versions, even only one GE, that could reside on a separate site with its own

Synchronization and Transport modules, while the LE and DF are sharing their SL and TL modules, as shown in Fig. 4.

c)   It is possible to have DF not associated with each version, e.g., have fewer DFs, or even only one DF. This DF can also be located on a separate site, as shown in Fig. 5.

d)   Generally, it is possible to construct specialized hardware units which contain separately the functions LE, DF, and GE. Only the LE module needs to have the capability of running a version.

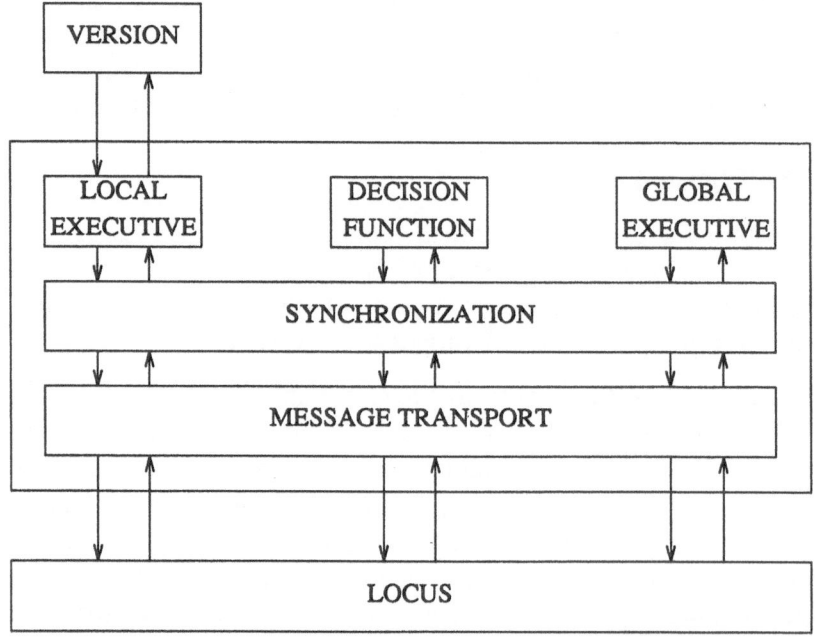

Fig. 3. The Standard DEDIX Configuration on One Site

The structure of DEDIX is adaptable for different applications. Depending on the reliability requirements and the reliability of the versions and the hardware, the decision algorithm and the reconfiguration possibilities, a special arrangement and solution can be chosen. The configurations of Fig. 4 and Fig. 5 have actually been used at UCLA for specific experiments.

3. Current Implementation of DEDIX

A prototype of DEDIX began operation in early 1985 [Avižienis 1985a]. It has been implemented using the C programming language and is running in

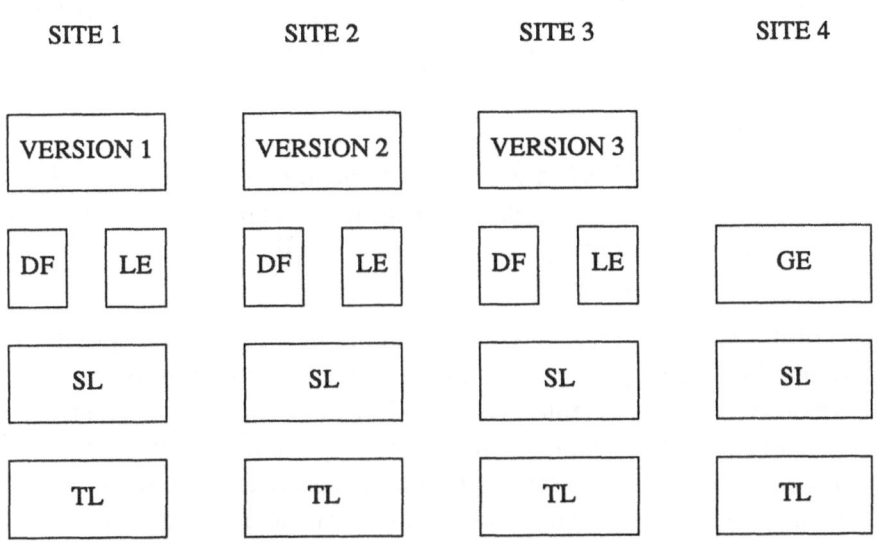

Fig. 4. 3 Sites with 3 Versions, LE, and DF, Site 4 with single GE

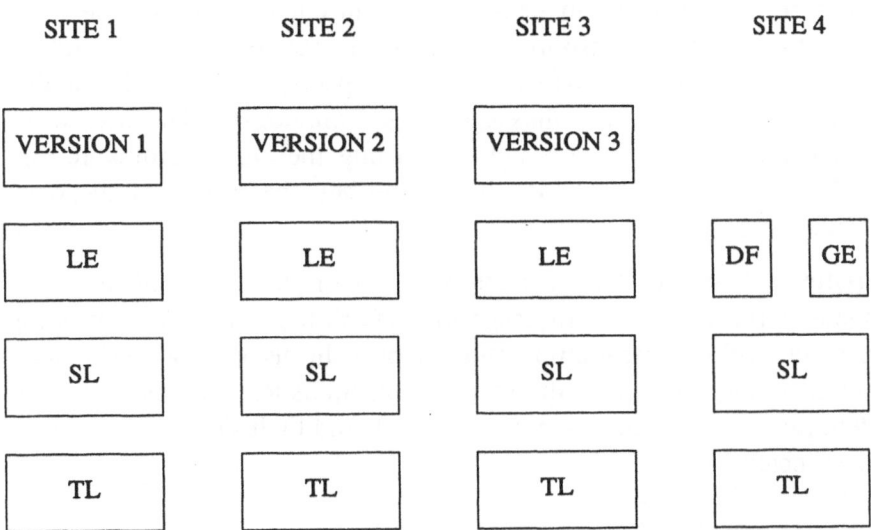

Fig. 5. 3 Sites with 3 Versions, Site 4 with DF and GE

the Locus environment at UCLA. Several modifications and refinements
have been incorporated in DEDIX since 1985, mostly to improve the speed
of N-version execution. The current standard realization has one LE, DF, and
GE for each version, running on the same site as the version. It is also

possible to execute multiple versions (and therefore also multiple DEDIX replicas) on one site - a single VAX 11/750 machine.

3.1 The User Interface

The user interface of DEDIX allows users to debug the system as well as the versions, to monitor the operations of the system, to apply stimuli to the system, and to collect data during experiments. Several commands are provided to the user, as discussed below.

Breakpoint. The *break* command enables the user to set breakpoints. At a breakpoint, DEDIX stops executing the versions and goes into the user interface, where the user can enter commands to examine the current system states, examine past execution history, or inject faults to the system. The *remove* command deletes breakpoints set by the break command. The *continue* command resumes execution of the versions at a breakpoint. The user may terminate execution using the *quit* command. The user is allowed to inject faults to the system by changing the system states, e.g., the cc-vector, by using the *modify* command.

Monitoring. The user can examine the current contents of the message passing through the Transport layer by using the *display* command. Since every message is logged, the user may also specify conditions in the display command to examine any message logged previously. The user can also examine the internal system states by using the *show* command, e.g., to examine the breakpoints which have been set, the results of the decision algorithm, etc.

Statistics collection. The user interface gathers data and collects statistics of the experiments. Every message that passes the transport layer is logged into a file with a time-stamp. This enables the user to do post-execution analysis or even to replay the experiment. Statistics such as elapsed time, system time, number of cc-points executed, and their decision outcomes are also collected.

3.2 The Version Interface

Cross-check functions. The programmer of a version must incorporate calls to the Cross-Check Functions (cc-functions) in order to make use of the support provided by DEDIX. These calls have to be included at logically identical cross-check points (cc-points) in the different versions which are going to communicate via DEDIX. Therefore, the cc-points have to be defined in the common specification of the versions. The cc-function calls have the following structure: *ccpoint (ccid, format, arguments)*. The parameter list is called *cc-vector*. The *ccid* is the identification of the cc-point. The *format* is a string of characters identifying the types of the variables contained in the arguments and the kind of decision algorithm which is to be applied to these variables. Possible variable types are character, integer, real, etc. The possible decision algorithms are described in Section 2.3.

The identity number of the cross-check point is passed on to the cc-function, to make sure that only information belonging to the same cc-point is compared by the Decision Function. Three different cc-function calls are possible: *ccinput* for input of data (instead of a standard input read statement), *ccoutput* for output of data (instead of a standard output write statement), and *ccpoint* for error recovery.

The ccinput call. The input to the versions is initiated via this call. The Decision Function checks whether all versions agree on the format of the input. In case of a single input, the input is distributed to all versions. In case of multiple versions of input, e.g. by redundant sensors, it is possible that either each version receives its related input, or that the Decision Function checks the inputs and chooses a consensus value for common distribution.

The ccoutput call. All outputs of the *N*-version software unit must be made via DEDIX and therefore must pass through the Decision Function. The results or data from the versions can be a collection of integers, character strings, and real numbers. Along with this data, a selection can be made on which kind of consensus decision shall be used for the outputs. Different consensus for different parts of the output may be specified.

The ccpoint call. This call is made if a cross-check between the versions is desired, but no input or output is required. Again, a decision is made on the version results, and the consensus result is passed back to the versions.

Recovery points. Complete error recovery of failed versions is performed at recovery points (r-points). Associated with each r-point in each version are: a recovery point id (*rp-id*), which uniquely identifies the r-point, and two exception handlers, the *state-input exception handler* and the *state-output exception handler*, that are required to input and to output the internal state of the version (*version state*) in a specified format. The r-point call has the following structure: *rpoint (rpid)*, where the rpid is the identification of the r-point.

4. Research Applications of DEDIX

The N-version software research at UCLA has two major long-term objectives:

(1)   to develop the principles of implementation and experimental evaluation of fault-tolerant N-version software units; and

(2)   to devise and evaluate supervisory systems for the execution of N-version software in various environments.

Both objectives are strongly supported by the experimental use and the continuing evolution of the DEDIX supervisory system. Some key aspects of the research applications of DEDIX are discussed below, and some recent specific results - in subsequent sections.

4.1 Implementation and Evaluation of N-Version Software

The N-version implementation studies that are supported by DEDIX address the problems of: 1) methods of specification, and the verification of specifications; 2) the assurance of independence of versions; 3) partitioning and matching, i.e., good choices of cc-points, r-points, and cc-vectors for a given problem; 4) the means to recover a failed version; 5) efficient methods of modification for N-version units; 6) evaluation of effectiveness and of cost; 7) the design of experiments.

Initial specification and partitioning. The most critical condition for the independence of design faults is the existence of a complete and accurate specification of the requirements that are to be met by the diverse designs. This is the "hard core" of this fault tolerance approach. Latent defects, such as inconsistencies, ambiguities, and omissions, in the specification are likely to bias otherwise entirely independent programming or logic design efforts toward related design faults. The most promising approach to the production of the initial specification is the use of formal, very-high-level specification languages that are discussed in Section 6. When such specifications are executable, they can be automatically tested for latent defects and serve as prototypes of the programs suitable for assessing the overall design. With this approach, perfection is required only at the highest level of specification; the rest of the design and implementation process as well as its tools are not required to be perfect, but only as good as possible within existing resource constraints and time limits.

The independent writing and subsequent comparisons of two specifications, using two formal languages, is the next step that is expected to increase the dependability of specifications beyond the present limits. Our current investigation of specification methods is discussed in Section 6. It is also important to note that a formal specification must specify the following application-specific features needed by *N*-version execution: 1) the initial state of the program; 2) the inputs to be received; 3) the location of cross-check and recovery points (partitioning into modules); 4) the content and format of the cross-check vector at each cc-point and r-point (outputs are included here); 5) the algorithms for internal checking and exception handling within each version; and 6) the time constraints to be observed by each program module.

Independence of version design efforts. The approach that is employed to attain independence of design faults in a set of *N* programs is maximal independence of design and implementation efforts. It calls for the use of diverse algorithms, programming languages, compilers, design tools, implementation techniques, test methods, etc. The second condition for independence is the employment of independent (noninteracting) programmers or designers, with diversity in their training and experience. Wide geographical dispersion and diverse ethnic backgrounds may also be desirable. DEDIX provides a suitable environment to study the effectiveness of efforts to attain diversity and independence of versions. Recent experimental results and plans for future experiments are reviewed in Section 5.

Recovery of failed versions. A problem area that has been addressed recently [Tso 1987a] is the recovery of a failed version in order to allow its continued participation in *N*-version execution. Since all versions are likely to contain design faults, it is critically important to recover versions as they fail rather than merely degrade to *N*-1 versions, then *N*-2 versions, and so on to shutdown. Recovery of a given version is difficult because the other (good) versions are not likely to have identical internal states; they may differ drastically in internal structure while satisfying the specification. The Community Error Recovery (CER) approach offers a systematic two-level method of forward recovery for failed versions [Tso 1987b]. Recent results of an experimental evaluation of CER using DEDIX are presented in Section 5.

Modification of *N*-version software. It is evident that the modification of software that exists in multiple versions is more difficult. The specification is expected to be sufficiently modular so that a given modification will affect only a few modules. The extent to which each module is affected can then be used to determine whether the existing versions should be modified according to a specification of change, or the existing versions should be discarded and new versions generated from the appropriately modified specification. DEDIX-based experiments are currently being planned to gain insights into the criteria to be used for a choice.

Assessment of effectiveness. The usefulness of the N-version approach depends on the validity of the conjecture that residual software faults in separate versions will cause very few, if any, similar errors at the same cc-points. Large-scale experiments need to be carried out in order to gain evidence on the nature of faults encountered in independently developed program versions. The ''mail order software'' approach offers significant promise to provide versions to be evaluated using DEDIX. An ''international mail order'' experiment is being planned, in which the members of research groups from several countries will use a formal specification to write software versions. It is expected that the software versions produced at widely separated locations, by programmers with different training and experience who use different programming languages, will contain substantial design diversity. In further experiments, it may be possible to utilize the rapidly growing population of free-lance programmers on a contractual basis to provide module versions at their own locations. This approach would avoid the need to concentrate programming specialists, have a low overhead cost, and readily allow for the withdrawal of individual programmers.

Cost investigations. The generation of N versions of a given program instead of a single one shows an immediate increase in the cost of software prior to the verification and validation phase. The question is whether the subsequent cost will be reduced because of the ability to employ two (or more) versions to attain mutual validation under operational conditions. Cost advantages may accrue because of 1) the faster operational deployment of new software; and 2) replacement of costly verification and validation tools and operations by a generic N-version environment (such as DEDIX) in which the versions validate each other while executing useful work. The loss of performance due to the presence of fault tolerance mechanisms, such as decision algorithms and recovery points also needs to be assessed.

Design of experiments. Several design issues of design diversity experiments need to be carefully resolved. They include: exploring different dimensions of diversity, incorporating efficient error detection and recovery algorithms, and avoiding commonalities in the design effort. The software versions produced in these experiments need to be subject to controlled conditions that approximate the development methodologies and environments used by advanced industrial facilities. There should be extensive logging of work periods and events such as error detection, specification questions and answers, and test suite execution. The experiment leaders need to provide a complete high-level, high-quality specification. At all stages, questions about the specifications are submitted by electronic mail, reviewed by the experiment leaders, and answered by electronic mail. The rule of "written communication only" makes it possible to control and analyze the information flow. The determination that a question revealed a flaw in the specifications causes changes to be broadcast to all programmers at all sites. The deliverable items include a design document, a series of compiled programs representing the results of the top down development at each abstraction layer, a test plan and test log, and the final program. The delivered software is then subjected to an adequate acceptance test to ensure its quality.

To measure the extent of design diversity and to assess potential reliability increases under large-scale, controlled experimental conditions, two major projects are underway: The NASA/Four-University experiment (initiated in the summer of 1985) and the UCLA/Honeywell Experiment (to be conducted during the second half of 1987). Descriptions of these two experiments are presented in the following Section 5.

4.2 Investigations of Supervisory Systems

The research concerned with N-version supervisory systems, as exemplified by DEDIX, deals with: 1) the functional structure of supervisors; 2) fault-tolerant supervisor implementation, including tolerance of design faults; 3) instrumentation to support N-version software experiments; 4) efficient implementation, including custom hardware architectures to support real-time execution; and 5) methods of supervisor evaluation.

***N*-version execution supervision and support.** Implementation of N-version fault-tolerant software requires special support mechanisms that need to be specified, implemented, and protected against failures due to physical or design faults. These mechanisms fall into two categories: those *specific* for the application program being implemented, and those that are *generic* for the N-version approach. The *specific* support is part of the version specification. The *generic* class of support mechanisms forms the N-version execution supervision environment that includes: 1) the decision algorithm; 2) assurance of input consistency; 3) interversion communication; 4) version synchronization and enforcement of timing constraints; 5) local supervision for each version; 6) the global executive and decision function for the recovery or shutdown of faulty versions; and 7) the user interface for observation, debugging, injection of stimuli, and data collection during N-

version execution of application programs. The nature of the generic support mechanisms has been illustrated in the discussions of the DEDIX *N*-version supervisor system that was described in preceding Sections 2 and 3. The continuing use of DEDIX leads to further insights that result in refinements and enhancements of DEDIX functional structure and its efficiency.

Protection of the supervisory environment. The success of design fault tolerance by means of *N*-version software depends on uninterrupted and fault-free service by the *N*-version supervision and support environment. Protection against physical faults is provided by the physical distribution of *N* versions on separate machines and by the implementation of fault-tolerant communication linkages. The SIFT system [Wensley 1978] and DEDIX are suitable examples in which the global executive is also protected by *N*-fold replication. The remaining problem is the protection against design faults that may exist in the support environment itself. This may be accomplished by *N*-fold diverse implementation of the supervisor. To explore the feasibility of this approach, the prototype DEDIX is currently undergoing formal specification. Subsequently, this specification will be used to generate diverse multiple versions of the DEDIX software to reside on separate physical nodes of the system. The practicality and efficiency of the approach remain to be determined. Some results are discussed in Section 6 of this paper.

Architectural support. Current system architectures were not conceived with the goal of *N*-version execution; therefore, they lack supporting instructions and other features that would make *N*-version software execution efficient. For example, the special instructions "take majority vote" and "check input consistency" would be very useful. The practical applicability on *N*-version software in safety-critical real-time applications hinges on the evolution of custom-tailored instruction sets and supporting architectures. The current DEDIX implementation supported by Locus is likely to be too slow for this purpose. Despite this limitation, the functional architecture of DEDIX can be used with faster transport service and faster scheduling policies in a real-time system, while Locus can be used to simulate real-time execution.

5. Testing Tools, Experience, and Results

5.1 Programs for Demonstration and Testing

This section describes the existing multiversion programs for DEDIX which were developed to test DEDIX and to demonstrate its capabilities. The most important characteristics of these application programs are also given. Most of the programs were written in the C programming language.

The *Airport Scheduler* simulates an airport database and is based on the specification used in [Kelly 1983]. Typical operations include: scheduling or canceling a flight, changing certain flight data (e.g. departure time), reserving a seat, and looking up information in the database. There exist three versions: one version implements the database using arrays, the second one uses linked lists. Both versions are written in C, and the third version is identical to the first, except that it is written in Pascal. These three programs demonstrate the concept of design diversity and are suitable to test DEDIX after modifications.

The programs *arcade* and *arc_io* can be used to test the implementation of the cross-check functions. A number of borderline cases are explored. Furthermore, calls which violate the specification of these functions are made in order to test the robustness of DEDIX. Examples are: calls without cross-check vectors, with inconsistent cc-point identifier, wrong format string, or inconsistencies between format string and cc-vector.

The program *cv* tests the implementation of the Decision Function. It reads test cases from a file, applies them to the decision algorithm, and stores the decision result in an output file. A file with some 100 standard test cases is available, as are the expected results from these test cases. Three versions exist which differ in that they read different files.

Name is a sample demonstration program that reads an integer, performs some "complicated" computations involving arrays of integer and real numbers, and that finally selects a string (name) from a table for display. Four versions exist: three mutated ones with different built-in data tables, and one that simulates an infinite loop.

The program *power* computes a to the power of b for two built-in numbers a and b. It is designed to exercise the number-handling features of DEDIX. Twenty different bugs (e.g. different numerical constants, typing errors, wrong use of cc-functions) have been injected into the program and can be invoked. Thus 2^{20} different versions can be generated. These versions serve as mutants for the validation of DEDIX by mutation testing.

Test is a test program to test the basic functions of DEDIX. It is similar to *arcade* and *arc_io*, but is better documented and tests more thoroughly. Standard test cases and their expected results are available. In addition, there exists a shell script that repeatedly executes these test cases in different configurations (single machine, distributed) or with different run time options of DEDIX.

Time is a demonstration program that reads the clock of the machine it is running on, converts it into a string and displays it. It also asks interactively whether that process should be repeated. *Time* demonstrates that diverse versions can have synchronized clocks. Naturally, this program is only useful when DEDIX is distributed, i.e. each version runs on a different machine.

Table 1 summarizes the characteristics of the above mentioned testing programs.

Table 1. Multiversion Programs for Demonstration and Testing Purposes

Name	Number of Versions	Lines of Code (approx.)	Language	Main Purpose
airport	3	470	C, Pascal	demonstration
arcade, arc_io	6	350	C	test - cc-functions
cv	3	100	C	test - voter
name	4	60	C	demonstration
power	20+	515	C	test - number handling, mutation testing
test	4	300	C	test - all basic functions
time	7	20	C	demonstration

5.2 Proper Specification and Testing of Fault Tolerance Mechanisms

It was noted earlier that the *N*-version error detection and recovery mechanisms for each version, including cross-check points and recovery points, need to be defined in the software requirement specification. To avoid restricting design diversity, the programmers may be given a choice where to place the cc-points in their programs. The sequence in which the cc-points occur and the variables involved should be specified, and it should be required that the variables of each cc-point be computed but not used before the cc-point is reached. The programmers are also required to use the (possibly modified) values returned by the DEDIX supervisor in all subsequent computations.

The acceptance test should adequately test the recovery capability. It should ensure that the cc-points are placed in the right sequence, and output values are checked in right places during the execution of each version. Possible design faults that are related to cc-points fall into two categories:

1. *Incorrectly located cc-points.* Some programmers might place cc-points before the final values are calculated. These cc-points are placed too early. Also, some versions might use computed values before passing them to the decision function. These cc-points will occur too late.

2. *Unused returned values.* This fault could occur when a version uses an internal variable in place of a state variable. The value of the internal variable is assigned to the state variable of the cc-vector before the cc-point is called, but subsequent computations are still based on the value of that internal variable.

These faults can be detected by specially designed tests. The output values should be checked at the cc-points. This will detect the incorrect placement of cc-points. Also, specific tests should be included that deliberately return new values to some cc-points. The results of the next cc-point should then be checked to verify that the returned values are actually used. These preparations are necessary for the proper execution of multi-version software in the DEDIX environment.

5.3 DEDIX in the NASA/Four-University Multiversion Software Experiment

The NASA Langley Research Center is sponsoring the NASA/Four-University experiment in fault-tolerant software which has been underway since 1984. During the summer of 1985, the NASA experiment employed 40 graduate students at four universities to design, code and document 20 diverse software versions of a program to manage redundancy and to compute accelerations for a redundant strapped down inertial measurement unit (RSDIMU). The analysis of this software currently engages researchers at six sites: UCLA, the University of Illinois at Urbana-Champaign, North Carolina State University, and the University of Virginia, as well as the Research Triangle Institute (RTI), and Charles River Analytics (CRA). Empirical results from this experiment will be jointly published by the cooperating institutions after the verification, certification, and final analysis phases are complete. While the joint results still await publication, some independent results from the UCLA effort have been reported in [Kelly

1986].

During the summer of 1985, each of the four universities employed ten graduate students to design, code and document five software versions in ten weeks. At the end of this effort, each of these 20 software versions was required to pass a preliminary acceptance test that used 75 test cases. At UCLA, a long and careful validation phase including extensive testing of the versions followed the 10-week software generation phase. During validation, many errors and ambiguities in the specifications and the software versions were revealed. The specifications were subsequently refined. The five UCLA versions have since been further debugged by the original programmers and have passed a final (UCLA) certification test that consisted of 200 random test cases, 55 hand-made test cases of special value test data and extremal value test data, and special test cases for verifying the recovery mechanism. The size of the five resulting software versions ranged from 1677 to 2794 lines of Pascal statements. The scope of this discussion is limited to the specific testing done at UCLA that employed DEDIX. The purpose of the tests was to evaluate the new CER forward recovery method [Tso 1987b]. Only the five certified versions from UCLA were used in these tests.

A Test Case Generator (TCG) was used throughout the evaluation of recovery to generate random test cases. After the TCG had generated the data for a test case, all five individual versions were executed consecutively, using the same input data. If a majority of similar results exists, they are used to decide the *reference* output which is further checked by the known TCG output values to ensure its consistency. At the same time, individual version failures are identified. This failure information is used to generate "interesting" 3-version combinations (triplets) using the assumption that all majority versions behave identically for that test case. This means that triplets with two good versions, such as (G1, G2, B), (G1, G3, B), and etc., are treated as one, i.e., (G, G, B), and many triplets can be eliminated from further testing. The *interesting* triplets are then executed in a three-version configuration under DEDIX supervision. The decision results are passed back to the failed versions for partial recovery at the cc-point level. Decision results of the triplets without recovery are obtained simply by comparing individual version outputs of the combinations. The decision results, both a) *without recovery* and b) *with recovery*, are then used to determine the effectiveness of the recovery. The process then is repeated for further test cases. A total of 200,000 test cases were employed in recovery evaluation.

5.3.1 Faults Discovered and Errors Observed During Testing

During the recovery evaluation process, several faults were found in the five UCLA certified versions. Table 2 lists these faults and their effects on the outputs, i.e., the errors seen at cc-points.

The fault ucla1-1 manifested itself during the testing because of the use of a Pascal compiler in the testing harness, while a Pascal interpreter was used in the program development and certification processes. Obviously, the interpreter initializes variables in a Pascal procedure, while the compiler does not. Since this fault failed the version more than half of the time, it was taken out in our evaluation. One of the display functions is to display the five most significant digits and the decimal point of a floating point number. Two versions failed to round the numbers correctly, although not in the same way. Both versions ucla3 and ucla4 made wrong system failure decisions, but for two different reasons. Thus the faults are different, but both versions produced coincident and identical errors at the cc-point for 96 out of the 200,000 test cases.

Failures of the individual versions. The result of a version running a test case is defined as erroneous if one or more of its output values (out of a total of 64 element values) differs from the *reference* values defined previously. We also say that the version *fails* on that test case. Table 3 shows the observed failures for the *individual* program versions for the 200,000 test cases, and their sizes in number of Pascal statements.

It must be noted that the failure probability depended very much on the test case generator, and on the range of variation ("skew") that is allowed when results are compared. We consider that the versions tested in this evaluation were under stress because the test cases were sampled randomly from the largest possible input space. In actual flight, extremal input data are much less likely to happen than routine data.

Coincident failures of the versions. Two versions are said to fail coincidently if they both fail (produce erroneous values of the same element) for the same test case. These coincident errors may be similar or distinct. It was observed that more than two versions did not fail for the same test case during the 200,000 test runs. There was one coincident error between ucla1 and ucla3, and there were 110 coincident errors between ucla3 and ucla4.

Similar errors of the versions. It should be noted that the results of the versions which fail coincidently may not be similar. *Similar results* are defined to be two or more results (good or erroneous) that are within the

Table 2. Characteristics of Discovered Faults

Label	Class	Fault	Error at cc-point
ucla1-1	incorrect algorithm	uninitialized variable	incorrect sensor status
ucla1-2	incorrect algorithm	bad display rounding	incorrect display
ucla1-3	incorrect algorithm	overflow handled incorrectly	incorrect display
ucla2	no fault discovered		
ucla3-1	spec mis-interpretation	individual instead of average slopes used	incorrect sensor status
ucla3-2	spec mis-interpretation	wrong frame of reference used	incorrect sensor status
ucla3-3	spec ambiguity	wrong system failure decision	incorrect system status
ucla3-4	incorrect algorithm	bad display rounding	incorrect display
ucla3-5	incorrect algorithm	overflow not handled	incorrect display
ucla4-1	spec ambiguity	wrong system failure decision	incorrect system status
ucla5	no fault discovered		

range of variation that is allowed by the decision algorithm. When two or more similar results are erroneous, they are called *similar errors* [Avižienis 1985b]. It was found that only ucla3 and ucla4 had similar errors which occurred for 96 test cases.

5.3.2 Results of CC-Point and R-Point Recovery

The effectiveness of partial recovery at cc-points was evaluated by

Table 3. Failures of Individual Versions

Version	Size	Number of Failures	Failure Probability
ucla1	2016	1	0.000005
ucla2	1685	0	0.000000
ucla3	1962	702	0.003510
ucla4	2794	283	0.001415
ucla5	1677	0	0.000000

comparing the *final decision* results of a 3-version RSDIMU software module (triplet) executed *without* the cc-point recovery provision and the one executed *with* the cc-point recovery provision. This was the first opportunity to perform cc-point recovery as an experiment.

The *final decision* of a triplet falls into five categories as shown in Table 4.

Table 4. Classification of Triplet Decisions

Final Decision	Individual Version Results			Explanation
GOOD3	G	G	G	All three results are good (G).
GOOD2	G	G	B	Only two results are good. The error (B) of the failed version is masked.
NOMAJ	B1	B2	G	All three results are different from each other. This decision is a fail-safe stop.
	B1	B2	B3	
BAD2	B	B	G	A similar error (B) occurs in two versions.
	B	B	B1	
BAD3	B	B	B	A similar error in all three versions.

Table 5 summarizes the consequences of including the cc-point recovery over the 200,000 test cases. Almost 90% of the changed decisions of the 3-version RSDIMU module are from GOOD2 to GOOD3, meaning that errors which occurred in a single version of the triplets had been recovered successfully by cc-point recovery. The improvement of a decision from GOOD2 to GOOD3 should not be diminished by the fact that the decision results of the two decisions are the same, and the change is only on the confidence level. This improvement makes the 3-version MVS system fully recovered and ready to tolerate another fault that may happen in the subsequent computations. There are 64 decisions in the GOOD3 → GOOD3 category although the triplets include one or two failed versions. This occurs because our analysis considers the System Status and Estimated Acceleration results only, and these failed versions were able to compute them correctly, but failed in the Display Driver.

Table 5. Consequences of CC-Point Recovery

Without recovery	With recovery	Triplets of 2 G and 1 B versions	Triplets of 1 G and 2 B versions	Total	Percent
GOOD3 → GOOD3		63	1	64	–
GOOD2 → GOOD3		923	3	926	89.6
NOMAJ → GOOD3		0	9	9	0.9
NOMAJ → BAD3		0	2	2	0.2
BAD2 → BAD3		0	96	96	9.3
Total number of changed 3-version RSDIMU module results				1033	100

There are nine triplets which had their decisions improved from NOMAJ to GOOD3. This improvement happens when two different versions fail at different cc-points. Without recovery, the triplet produced a NOMAJ decision; with recovery, it first recovered a failed version at an earlier cc-point, then the fully recovered triplet recovered another failure later. Since the RSDIMU module has only five computations, and most of the observed errors occurred after the second one, the case in which a fully recovered triplet recovered from a second fault happened rather rarely.

In the 96 triplets that had their decisions changed from BAD2 to BAD3 the good version was forced to fail in the same way by an attempted recovery. However, similar errors already existed in a majority of versions, and the MVS system is assumed to fail, in either case. The two triplets that have their decisions changed from NOMAJ to BAD3 are dangerous because the 3-version MVS system has been changed from a fail-safe state to an unsafe state.

The most frequent similar errors observed during the testing are due to the case in which both versions ucla3 and ucla4 declare that the RSDIMU system failed. This decision sets all sensor status to non-operative and the estimated accelerations to zero. The program faults (ucla3-3 and ucla4-1 in

Table 2) are due to extra checks on conditions that should not happen according to the original RSDIMU specification. This specification was changed during the course of program development and certification. However, it should be noted that such outputs lead to a fail-safe response of shutting down the system in the RSDIMU application. Detailed discussion of the results appears in [Tso 1987b].

Recovery points were not specified in the RSDIMU specification. However, a new program can be easily composed in which the RSDIMU module is the first module, with an auxiliary (AUX) module added. Then a recovery point is inserted between them. The AUX module contains nothing but a new cc-point used to check if the AUX module is indeed reached and started with a correct version state. The version state at the beginning of the AUX module was defined to be the collection of all the eleven output variables and of two other variables in the RSDIMU module found to be common to all the versions. One of them is the id number of the failed face, and the other is the threshold that determines a sensor failure. With the version state defined, state input and output exception handlers were implemented and used by all five versions.

DEDIX was used for testing because recovery at the recovery point level requires a sophisticated N-version supervisor to keep track of errors detected at the cc-points, to invoke the exception handlers, and to restart an aborted version. All the test cases that caused some versions to fail during previous cc-point recovery testing were used to test triplets of the instrumented programs for r-point recovery. The evaluation is similar to the previous one that considered RSDIMU system improvement with cc-point recovery. The previous evaluation examined the final results (System Status and Estimated Acceleration) of a 3-version RSDIMU module. In this evaluation we examined the version state after the recovery point.

The results of all possible consequences of a DEDIX test run executing a triplet of instrumented versions (containing either one, or two bad versions) are shown in Table 6. Each version consists of the RSDIMU module, an r-point, and the AUX module. Table 6 shows that for triplets with only one bad version, 983 of the 986 version states of the bad versions were recovered correctly at the recovery point, and there were 3 cases in which DEDIX gave the "No Majority" decision because of disagreement in comparing version states. The good versions used to form the triplet were the first two good versions chosen in the order of their version identifiers, therefore they always were different versions. It was found in those 3 test runs that although ucla1 had produced good outputs at the cc-points in the RSDIMU

module, in fact it had an erroneous internal state that was revealed by the
two additional variables included in the version state specification.

All 98 triplets of one good and two bad versions that had produced BAD3
decisions with cc-point recovery (see Table 5) had the "No Majority" decision while comparing the version states. This happened because the two versions both had incorrectly concluded for different reasons that the RSDIMU
module failed, and thus had produced *similar errors at the cc-point that
were due to different faults*. However, the two common non-output variables differed and therefore a BAD3 majority decision was avoided. This
occurrence shows that r-point checking is more effective than only cc-point
checking since the BAD3 cc-point decisions were properly detected at the r-
point.

Table 6: Consequences of R-Point Recovery

Possible Consequence	Triplets with 1 B Version	Triplets with 2 B Versions	Total
No majority at the cc-points in the RSDIMU module	0	0	0
No majority in comparing the r-point ids	0	0	0
No majority in comparing the version states at r-point	3	108	111
GOOD3 decision at the cc-point in the AUX module	983	3	986
GOOD2 decision at the cc-point in the AUX module	0	0	0
NOMAJ decision at the cc-point in the AUX module	0	0	0
BAD2 decision at the cc-point in the AUX module	0	0	0
BAD3 decision at cc-point in the AUX module	0	0	0

There are also three test runs of triplets with 2 bad versions that produced the
GOOD3 decision at the last cc-point. This happened because the errors of

the two bad versions had occurred at different cc-points and were successfully recovered by cc-point recovery (Table 5).

Since control flow errors were not observed in the 200,000 test runs, more testing was conducted through error seeding. The goal was to verify the effectiveness of the restart mechanism of the r-point. Faults of the following two categories were seeded: 1) *Program exceptions*, such as "division by zero" and "index out of range," and 2) *control flow faults*, such as "infinite loop" and "incorrect branching" that lead to some cc-point being incorrectly called or skipped.

Most of the faults that were seeded into the versions were chosen from faults that were eliminated during the certification process. Testing was conducted with triplets consisting of two good versions combined with a version with a seeded fault. It was found that in all the hundred different test runs that were performed, the failed versions were restarted with a correct version state after the recovery point.

5.4 The UCLA/Honeywell Fault-Tolerant Software Experiment

To gain further insights into the effectiveness and methodology of applying multi-version software systems, UCLA and the Honeywell - Sperry Commercial Flight Systems Division have agreed to conduct a joint study of multi-version software design during the second half of 1987. The application is the digital flight control system for future commercial airliners, as exemplified by the system being developed by Honeywell for potential use in the McDonnel-Douglas MD-11 aircraft.

The objectives of the UCLA/Honeywell project to study the *N*-version flight control system design are as follows:

- To conduct studies and experiments related as closely as practical to the industrial environment in terms of procedures and types of problems.

- To develop a practical and effective set of ground rules for multi-version software development in an industry environment. These ground rules will be directed toward the elimination of significant similar errors in the versions.

- To estimate the effectiveness of multi-version software in an industrial environment of a specified type.

The extent and purpose of the multiversion software is:

(a)   The software provides automatic pitch control of commercial aircraft during final approach.

(b)   The elements of the control loop are control law, airplane, sensors mounted on airplane, landing geometry, and wind disturbances.

(c)   Independent two-programmer teams will program the control laws, based on a software requirements document, i.e., the software specification.

(d)   The aircraft and wind turbulence are to be modeled on VAX machines. The operation of flight simulation will be monitored by DEDIX to observe the execution of a multi-version software system.

For the software development phase, six teams of two graduate student programmers each will work in the software development phase for 12 weeks during the summer of 1987. Software engineering techniques to build high quality software will be strictly followed. The six teams will be coordinated by the UCLA research team, using an electronic mail communication facility. A standard industrial design review, code review, and a test review will be conducted. The expected length of code produced in this software development phase should exceed 2000 lines.

Several dimensions of design diversity have been considered for achieving diversity among these programs. The attention will be focused on the use of different programming languages and their effects on the diversity in multi-version software. The languages are C, Pascal, Modula-2, Ada, Lisp, and Prolog.

In the evaluation of resulting multi-version software systems, closed loop testing of multiple executions with random inputs will be conducted. Millions of test runs will be executed in the DEDIX environment for suitable aircraft control and flight simulation. Statistical data related to execution of multi-version software systems will be gathered for the evaluation of the effectiveness of DEDIX.

6. Specification Issues

Significant progress has occurred in the development of formal specification languages, methods, and tools since our previous experiments [Kelly 1983, Avižienis 1984]. Our current goal is to compare and assess the applicability to practical use by application programmers of several formal program specification methods. The leading candidates are:

(1)   The Larch family of specification languages developed at MIT and the DEC Western Research Center [Guttag 1985];

(2)   The OBJ specification language developed at UCLA and SRI International [Goguen 1979];

(3)   The Ina Jo specification language developed at SDC [Locasso 1980];

(4)   The executable specification language "PAISLey" developed at AT&T Bell Laboratories [Zave 1986].

The study focuses on the assessment of the following aspects of the specification languages: (1) The purpose and scope, i.e., the problem domain; (2) completeness of development; (3) quality and extent of documentation; (4) existence of support tools and environments; (5) executability and suitability for rapid prototyping; (6) provisions of notation to express timing constraints and concurrency; (7) methods of specification for exception handling; and (8) extensibility to specify the special attributes of fault-tolerant multi-version software.

The goal of the study is the selection of two or more specification languages for the subsequent experimental assessment of their applicability in the design of fault-tolerant multi-version software. Two major elements of the experiment will be:

(1)   the concurrent mutual verification of two specifications by symbolic execution and mutual interplay;

(2)   an assessment of the practical applicability of the specifications, as they are used by application programmers in an N-version software experiment.

The next step in DEDIX development will be a formal specification of parts of the current DEDIX prototype (implemented in C): the Synchronization Layer, the Decision Function, and the Local and Global Executives. Among them, the Larch specifications of the Decision Function [Tai 1986] and of the Synchronization Layer have been constructed. The specification will provide an executable prototype of the DEDIX supervisory system. This functional specification should allow not only the migration to real-time systems, but also the use of multi-version software techniques for the fault-tolerance mechanisms of DEDIX themselves. The goal is a DEDIX system that supports design diversity in application programs and which is itself diverse in design at each site.

Independent specifications of some DEDIX system modules in two or more formal languages will serve to compare the merits of the methods. Further research is planned in the application of *dual* diverse formal specifications to eliminate similar errors that are traceable to specification faults and to increase the dependability of the specifications.

7. Other Current Research Activities

7.1 Improvement of DEDIX

This section discusses some observed deficiencies of DEDIX and offers some thoughts about improving them. Current activities are also mentioned, where appropriate.

The most visible shortcoming of DEDIX is the execution overhead which results in rather long waiting times for the user. There are two possible ways to improve the situation: one is to create a "custom DEDIX" which is tailored to a specific application. Functions that are not needed can be removed, and the versions can be compiled into DEDIX instead of creating another process for each version to be executed. That reduces greatly the amount of time spent with interprocess communication. The second approach is to look for more efficient implementations of these parts of DEDIX that are used most. Due to the layered design it should be relatively easy to replace a layer with a more efficient one, without affecting the others. Since most time is spent on message passing, an investigation of a more efficient implementation of the transport layer is under way.

Another observation is that DEDIX supports only standard input and output. Thus the ability to manipulate files is limited to redirecting the input and the output. Of course, it is possible for a version to use all the file manipulating operations that are provided by the operating system. However, the checking and correcting facilities of DEDIX would be essentially bypassed in this case. Furthermore, different versions may not read and/or write the same file(s) because that would result in an almost certainly unpredictable interaction between the versions. A solution would be to provide cross-check functions for file I/O, similar to those now provided for standard I/O. However, the following considerations lead to the conclusion that this is not too urgent: diverse software is likely only to be required and applied in systems with ultra-high reliability requirements, e.g. autopilots, flight control systems, air traffic control systems, or nuclear power plant control systems. Systems of these kinds are usually computation intensive, rather than data and I/O intensive. Thus it can be expected that it will be sufficient to support standard I/O for most of these systems.

Furthermore, the versions are limited to sequential programs all of which must execute all the specified cross-check points in the same order. In many cases the sequence of cc-points is given by the data flow of the computation to be performed. However, in case there are several independent submodules which could be executed in any order, a specific sequence of these independent computations has to be specified and all versions have to adhere to it. This, to a certain extent, limits the degree of diversity that could be achieved. Presently, neither a study examining whether this restriction is a severe one or not, nor a method to overcome it, exist. Of course, it is easy if DEDIX only *observes* the computed results, without trying to correct them – we just postpone the analysis until all versions have terminated.

7.2 Extension of DEDIX Capabilities

Byzantine faults [Lamport 1982] are defined as faulty behavior that may prevent agreement about the current (global) system state among the sites of a distributed system. Examples of such behavior include:

- sending more or fewer messages than required to by the protocol,
- sending messages too late or too early,
- sending different (inconsistent) information to different sites, or
- maliciously cooperating with another malicious site.

The Synchronization Layer of DEDIX provides considerable protection against the first two examples of faulty behavior. Since the topology of the current implementation is a ring structure, a site cannot send different information to different sites, but it can alter the information that it is supposed to forward. At the present time, DEDIX does not deal with other types of Byzantine (malicious) faults. Methods to tolerate them are known [Lamport 1982] and could be included in the Transport layer. A study is currently in progress that will provide some experimental data on the time and complexity overhead of these methods.

In order to build an elegant, highly reliable system which is tolerant to both hardware and software design faults, a study is in progress how to build a DEDIX system on top of a XEROX Worm environment [Shoch 1982]. The key idea is that the Worms bring a special philosophy to building distributed, fault-tolerant systems. This philosophy gives each individual unit a high degree of autonomy and a desire to complete its task and to take an active part in the activity of the whole system; and further, takes a network service approach to the resources available in the system.

8. Conclusion

This paper has presented an overview of a major effort to develop a research environment for software design diversity research at the UCLA Dependable Computing and Fault-Tolerant Systems Laboratory. The complete DEDIX prototype has been implemented, and it is being used to execute, test, and evaluate multiversion software. Some new research efforts also have been initiated.

Acknowledgment

The research described in this paper has been supported by a grant from the Advanced Computer Science program of the U.S. Federal Aviation Administration, by NASA contract NAG1-512, and by NSF grant MCS 81-21696. Professor Algirdas Avižienis, Director of the UCLA Dependable Computing and Fault-Tolerant Systems Laboratory, has served as Principal Investigator since the inception of the DEDIX project.

The original concept and implementation of DEDIX, as described in [Avižienis 1985a], has benefited from major contributions of several individuals who were visiting researchers at UCLA in the 1983-85 period. A large part of the DEDIX implementation is due to Lorenzo Strigini, who is currently at the IEI-CNR, Pisa, Italy. The communication and synchronization protocols are the contribution of Per Gunningberg, presently at the Swedish Institute of Computer Science, Stockholm, Sweden. The original decision function was designed and implemented by Pascal Traverse, now at Aerospatiale, Toulouse, France. John P. J. Kelly, now at the University of California, Santa Barbara, contributed extensive consultation on issues of experimentation and software engineering.

All authors of this paper are presently engaged in DEDIX-related research activities at UCLA, except as noted next. Udo Voges edited the first draft of this paper prior to returning to his permanent position at the Kernforschungszentrum Karlsruhe, Federal Republic of Germany. Kam Sing Tso has recently assumed a position at the Jet Propulsion Laboratory, Pasadena, California, U.S.A. We also wish to acknowledge the idea of fusing the XEROX Worm and DEDIX concepts, which is due to Nick Lai, a staff member at the UCLA Center for Experimental Computer Science.

References

[Anderson 1981]   T. Anderson and P. A. Lee, "Fault Tolerance: Principles and Practice," *Prentice Hall International,* London, England, 1981.

[Anderson 1985]   T. Anderson, P. A. Barrett, D. N. Halliwell, D. N. and M. R. Moulding, "An Evaluation of Software Fault Tolerance in a Practical System," *Digest of FTCS-15, the 15th International Symposium on Fault-Tolerant Computing*, Ann Arbor, Michigan, June 1985, pp. 140-145.

[Avižienis 1975]   A. Avižienis, "Fault-Tolerant and Fault-Intolerance: Complementary Approaches to Reliable Computing," *Proceedings of the 1975 International Conference on Reliable Software*, Los Angeles, April 1975, pp. 458-464.

[Avižienis 1977]   A. Avižienis and L. Chen, "On the Implementation of N-version Programming for Software Fault Tolerance During Execution," *Proceedings of the 1st IEEE-CS International Computer Software and Applications Conference (COMPSAC 77)*, Chicago, November 1977, pp. 149-155.

[Avižienis 1984]   A. Avižienis and J. P. J. Kelly, "Fault-Tolerance by Design Diversity: Concepts and Experiments," *Computer*, Vol. 17, No. 8, August 1984, pp. 67-80.

[Avižienis 1985a]   A. Avižienis, P. Gunningberg, J. P. J. Kelly, L. Strigini, P. J. Traverse, K. S. Tso, and U. Voges, "The UCLA DEDIX System: A Distributed Testbed for Multiple-Version Software," *Digest of FTCS-15, the 15th International Symposium on Fault-Tolerant Computing*, Ann Arbor, Michigan, June 1985, pp. 126-134.

[Avižienis 1985b]   A. Avižienis, "The N-Version Approach to Fault-Tolerant Software," *IEEE Transactions on Software Engineering*, Vol. SE-11, No. 12, December 1985, pp. 1491-1501.

[Chen 1978]   L. Chen and A. Avižienis, "N-version Programming: A Fault Tolerance Approach to Reliability of Software Operation," *Digest of FTCS-8, the 8th International Symposium on Fault-Tolerant Computing*, Toulouse, France, June 1978, pp. 3-9.

[Cristian 1982]   F. Cristian, "Exception Handling and Software Fault Tolerance," *IEEE Transactions on Computers*, Vol. C-31, No. 6, June 1982, pp. 531-540.

[Goguen 1979]   J. A. Goguen and J. J. Tardo, "An Introduction to OBJ: A Language for Writing and Testing Formal Algebraic Program Specifications," *Proceedings of the Conference on the Specification of Reliable Software*, Cambridge, MA, April 1979, pp. 170-189.

[Gunningberg 1985]   P. Gunningberg and B. Pehrson, "Specification and Verification of a Synchronization Protocol for Comparison of Results," *Digest of FTCS-15, the 15th International Symposium on Fault-Tolerant Computing*, Ann Arbor, Michigan, June 1985, pp. 172-177.

[Guttag 1985]   J. V. Guttag, J. J. Horning and J. M. Wing, "Larch in Five Easy Pieces," *Digital Equipment Corporation Systems Research Center, Report No. 5*, Palo Alto, California, July 24, 1985.

[Kelly 1983]   J. P. J. Kelly and A. Avižienis, "A Specification-Oriented Multi-Version Software Experiment," *Digest of FTCS-13, the 13th International Symposium on Fault-Tolerant Computing*, Milano, Italy, June 1983, pp. 120-126.

[Kelly 1986]   J. P. J. Kelly, A. Avižienis, B. T. Ulery, B. J. Swain, R. T. Lyu, A. Tai and K. S. Tso, "Multi-Version Software Development," *Proceedings of the IFAC Workshop SAFECOMP 86*, Sarlat, France, October 1986, pp. 43-49.

[Kim 1984]   K. H. Kim, "Distributed Execution of Recovery Blocks: An Approach to Uniform Treatment of Hardware and Software Faults," *Proceedings of the 4th IEEE International Conference on Distributed Computing* Systems, San Francisco, California, May 1984, pp. 526-532.

[Lamport 1982]   L. Lamport, R. Shostak and M. Pease, "The Byzantine Generals Problem," *ACM Transactions on Programming Languages and Systems,* Vol. 4, No. 3, July 1982, pp. 382-401.

[Locasso 1980]   R. Locasso, J. Scheid, V. Schorre and P. Eggert, "The Ina Jo Specification Language Reference Manual," *System Development Corp., Tech. Rep. TM-6889/000/01,* Santa Monica, California, November 1980.

[Melliar-Smith 1982]   P. M. Melliar-Smith and R. L. Schwartz, "Formal Specification and Mechanical Verification of SIFT: A Fault-Tolerant Flight Control System," *IEEE Transactions on Computers,* Vol. C-31, No. 7, July 1982, pp. 616-630.

[Popek 1981]   G. Popek, B. Walker, J. Chow, D. Edwards, C. Kline, G. Rudisin and G. Thiel, "LOCUS: A Network Transparent, High Reliability Distributed System," *Proceedings of the 8th Symposium on Operating Systems Principles,* Pacific Grove, California, December 1981, pp. 169-177.

[Ramamoorthy 1981]   C. V. Ramamoorthy, Y. Mok, F. Bastani, G. Chin and K. Suzuki, "Application of a Methodology for the Development and Validation of Reliable Process Control Software," *IEEE Transactions on Software Engineering,* Vol. SE-7, No. 6, November 1981, pp. 537-555.

[Shoch 1982]   J. F. Shoch and J. A. Jupp, "The 'Worm' Programs – Early Experience with a Distributed Computation," *Communications of the ACM,* Vol. 25, No. 3, March 1982, pp. 172-180.

[Tai 1986]   A. T. Tai, "A Study of the Application of Formal Specification for Fault-Tolerant Software," *M.S. thesis,* UCLA Computer Science Department, Los Angeles, California, June 1986.

[Tso 1987a]   K. S. Tso, "Recovery and Reconfiguration in Multi-Version Software," *Ph.D. dissertation,* UCLA Computer Science Department, University of California, Los Angeles, March 1987; also *Technical Report No. CSD-870013,* March 1987.

[Tso 1987b]   K. S. Tso and A. Avižienis, "Community Error Recovery in N-Version Software: A Design Study with Experimentation," *Digest of FTCS-17, the 17th International Symposium on Fault-Tolerant Computing,* Pittsburgh, Pennsylvania, July 1987.

[Voges 1982]   U. Voges, F. Fetsch and L. Gmeiner, "Use of Microprocessors in a Safety-Oriented Reactor Shut-Down System," *Proceedings EUROCON,* Lyngby, Denmark, June 1982, pp. 493-497.

[Wensley 1978]   J. H. Wensley, L. Lamport, J. Goldberg, M. W. Green, K. N. Levitt, P. M. Melliar-Smith, R. E. Shostak and C. B. Weinstock, "SIFT: Design and Analysis of a Fault-Tolerant Computer for Aircraft Control," *Proceedings of the IEEE,* Vol. 66, No. 10, October 1978, pp. 1240-1255.

[Zave 1986]   P. Zave and W. Schell, "Salient Features of an Executable Specification Language and Its Environment," *IEEE Transaction on Software Engineering,* Vol. SE-12, No. 2, February 1986, pp. 312-325.

6

Modelling Issues

The modelling of fault tolerant systems, especially those incorporating recovery blocks or multi-version software, is an active research issue. Some attempts in this direction have been made; examples are [Hecht 1979, Grnarov 1980, Bhargava 1981, Soneru 1981, Wei 1981, Migneault 1982, Laprie 1984, Eckhardt 1985, Scott 1987, Littlewood 1987].

In 1986, some thirty invited experts met in Badgastein, Austria, for a three day workshop to discuss dependability modelling of fault-tolerant software. The discussions ranged over single version programs, diverse versions and full diversity with adjudication (fault-tolerance) and included models and metrics, experiments and real-world practice. In the final sessions, recommendations for future research directions were given.

The following chapter summarises these conclusions.

References

[Bhargava 1981]   B. Bhargava, "Software Reliability in Real-Time Systems," in *Proc. National Computer Conference,* Chicago: 1981, pp. 297-309.

[Eckhardt 1985]   D. E. Eckhardt and L. D. Lee, "A Theoretical Basis for the Analysis of Multiversion Software Subject to Coincident Errors," *IEEE Trans. on Software Engineering,* Vol. SE-11, No. 12, December 1985, pp. 1511-1517.

[Grnarov 1980]   A. Grnarov, J. Arlat, and A. Avižienis, "On the Performance of Software Fault-Tolerance Strategies," in *Proc. 10th Intern. Symp. on Fault-Tolerant Computing FTCS' 10,* Kyoto, Japan: 1-3 October 1980, pp. 251-253.

[Hecht 1979]   H. Hecht, "Fault-Tolerant Software," *IEEE Trans. Reliability,* Vol. R-28, No. 3, August 1979, pp. 227-232.

[Laprie 1984]   J.-C. Laprie, "Dependability Evaluation of Software Systems in Operation," *IEEE Trans. on Software Engineering,* Vol. SE-10, No. 6, November 1984, pp. 701-714.

[Littlewood 1987]   B. Littlewood and D. R. Miller, "A Conceptual Model of the Effect of Diverse Methodologies on Coincident Failures in Multi-Version Software," in

Proc. 3rd Intern. Conf. Fault-Tolerant Computing Systems, Bremerhaven, Germany: 9-11 September 1987, pp. 263-272.

[Migneault 1982]   G. E. Migneault, ''The Cost of Software Fault Tolerance,'' in *Proc. AGARD Symposium on Software for Avionics, CPP-330,* The Hague, The Netherlands: 1982, pp. 37.1-37.8.

[Scott 1987]   R. K. Scott, J. W. Gault, and D. F. McAllister, ''Fault-Tolerant Software Reliability Modeling,'' *IEEE Trans. on Software Engineering,* Vol. SE-13, No. 5, May 1987, pp. 582-592.

[Soneru 1981]   M. D. Soneru, ''A Methodology for the Design and Analysis of Fault-Tolerant Operating Systems,'' Illinois Institute of Technology, Chicago, IL, USA, Tech. Rep. PhD Dissertation, May 1981.

[Wei 1981]   A. Y.-W. Wei, ''Real-Time Programming with Fault Tolerance,'' University of Illinois, Urbana, IL, USA, Tech. Rep. PhD Dissertation, 1981.

Reliability Modelling
for Fault-Tolerant Software

Report on a Workshop
Held in Badgastein, Austria, July 1986

B. Littlewood
Centre for Software Reliability
City University
Northampton Square
London, EC1V 0HB

T. Anderson
Centre for Software Reliability
The Computing Laboratory
University of Newcastle upon Tyne, NE1 7RU

1. Introduction: Aims of the Workshop

The overall aims of the workshop were to evaluate the current state of the art and, through this evaluation, lay the groundwork for future research directions.

In the early discussions of the organising committee it was agreed that the major output would be a document listing recommendations for research in this area. To this end it was felt to be important that we avoided the "conference" format of formal presentation of research papers followed by minimal discussion. Instead, we encouraged extensive discussion on focussed topics with only short motivating presentations. This approach was successful largely as a result of restricting attendance to a small group of invited participants, carefully selected for their relevant experience and expertise.

The workshop comprised five half-day sessions. The first three were on the themes **single version programs**, **diverse versions**, and **diversity with adjudication**. The organising committee decided that it would be useful to separate the issues of diversity and adjudication. Similarly, it was thought that an understanding of each of these stages must necessarily rest upon an understanding of single version programming.

In the event, our recommendations (which are detailed in the next section) did not completely fit into these neat divisions. In particular, specification issues were recognised to be of fundamental importance and so received special attention.

As well as the above classification of the problem, we invited participants to think about the key issue of modelling: the relationship between theory and practice. We hoped that, on the one hand, the workshop would provide recommendations for experiments to investigate theoretical hypotheses. On the other hand, modelling directions would arise from discussion of the results of experiments and practice.

2. Recommendations for Research

The final session of the workshop began with a strong affirmation of the potential value of fault-tolerant software approaches. The meeting recognised that much further work is needed in order to assess the magnitude of the dependability improvement attainable by means of fault-tolerance, and detailed recommendations for specific research are reported in the following. It was stated frequently during the meeting that the association of fault tolerance solely with ultra-high reliability applications can be misleading; it is just as likely to find use in more mundane contexts.

Apart from the detailed research recommendations which follow, there was agreement that an interaction between modelling and experimentation can help us towards an understanding of the problems, as well as yielding specific numerical measures.

Detailed recommendations follow. In many cases, explanatory comments follow the recommendations; in others the recommendations are felt to be self-explanatory. Absence of these comments does not imply that the recommendations are less important.

2.1 Modelling the Single Version Failure Process

During discussion, it became clear that there remained difficulties of definition and that these were giving rise to misunderstanding. Agreement here clearly requires better understanding of the failure process of single programs. In addition, it is obvious that an understanding of fault-tolerant diverse systems must depend on an understanding of the behaviour of the component versions from which they are constructed. This seems particularly important for safety critical applications, although it should not be thought that this is the only context in which fault-tolerance is likely to be relevant.

The following areas need investigation:

Formalisation of the input space model

In this model, execution of the program is represented by successive selection of inputs (points) in an input space. Execution is thus represented as a trajectory in this space, and faults are represented by subsets. We currently know very little about the nature of either the trajectories or of the fault sets, yet an understanding of the **failure process** depends on the interaction between these. We need to answer certain topological questions concerning "closeness" of inputs identified with a particular fault, and "closeness" of successively selected inputs during operational execution. Are trajectories connected paths in a suitably chosen topology? Are faults connected sets? Can random walk models avoid these difficulties? What are the implications for the failure process? For example, can we expect error bursts? It seems essential that these investigations be supported by controlled experiments.

Models for testing, repair and maintenance

Models to incorporate subjective and extraneous knowledge

Most of the current reliability models for single programs are reliability growth models which use failure data obtained during a period when faults are being rectified. They do not take any account of other information which is usually available, for example, the development methodology, characteristics of the program, and even expert judgements of the developers. Much of this kind of information is by its very nature "soft": it concerns influential rather than determining factors. Perhaps Bayesian statistical approaches, particularly methods for elicitation of opinion from experts, would be useful.

2.2 Single Version Software Testing and Verification

Current methods of testing software do not directly address the problem of reliability achievement. We do not know whether these methods are in any sense optimal for the achievement of reliability, nor do we know how to obtain credible figures for the ultimate user-perceived reliability from such testing. Similar observations are pertinent for other means of achieving and assessing software, such as the currently fashionable "formal methods".

The following issues must be addressed:

Experimental evaluation of the reliability benefits from different verification and testing strategies

Development of test beds for software testing

Investigation of "accelerated testing" techniques for software

It is often argued that testing in a user environment (so called "random testing") is very inefficient. We need to determine whether this is the case, and if so, seek methods for improving efficiency whilst retaining the important advantage of random testing: that it generates data of a form suitable for the calculation of reliability measures.

Development of an assertion-based testing methodology

Relevant issues here are the use of AI techniques for assertion writing, an exploration of data dependencies in general testing strategies, and the use of "watchdog" timers to execute assertions.

Investigation of applicability of hardware ideas

Examples are: use of graph models of programs to analyse (by fault simulation) the error environments encountered during actual system operation; test generation using the D-algorithm for combinational logic circuits.

2.3 General Issues of Diversity

The success of the fault-tolerant approach clearly depends on the degree to which we can achieve *diversity*. However, it is important, even in such an informal assertion, to distinguish carefully between the different aspects of diversity: diversity of the development processes used to obtain the versions, diversity of the actual version implementations, and diversity of the failure behaviour of these versions. In each case we need to know what we **mean** by diversity, how to **achieve** it, and how to **measure** the extent of our achievement.

Understanding of the diversity of the development processes

A clear understanding of what we mean by this kind of diversity is currently lacking, although most practitioners would be able to make subjective judgements. We need a definition of diversity which opens up the possibility of metrication. Only then shall we be in a position to identify highly orthogonal processes (for example in language, data structure, algorithm) which might be expected to produce the required diversity of version and version behaviour. It is possible, even likely, that process diversity is different for different fault classes. Finally, we need theoretical models which relate process to product: for example regarding the versions as random variable outputs of the development process.

Understanding of the diversity of particular implemented versions

Again we need understanding and agreed definitions: what does it mean to assert that a set of versions is diverse, that one set is "more" diverse than another? Version diversity must be defined in such a way that we can obtain a metric or metrics for "degree of diversity". This might, for example, be based on the structures of the versions. We need to know how version diversity depends on the diversity of the processes used to generate the versions; this needs modelling and experimentation. We also need to understand how diversity depends on other factors, for example, to what extent do synchronisation points diminish diversity? How does maintenance affect diversity?

Understanding of the diversity of output from different versions

Once again we need a definition and measures of degree of diversity. Simple early models have often used statistical independence of failure behaviour as a goal. However, it has long been recognised that such independence would be difficult, if not impossible, to attain. On the other hand, it **may** be possible to do better than independence, at least for certain classes of faults, by using "complementary" versions. In each case we need to measure what is achieved and thereby investigate the relationship between diversity of version output and diversity of processes and versions. Experiments and models are needed for this.

2.4 Issues of Fault-Tolerance; Diversity with Adjudication

An understanding of fault-tolerant systems requires more than an understanding of single version and diverse version software. It needs, in addition:

Understanding of the adjudication process

We need an understanding of different adjudication mechanisms, for example: averaging, consensus, acceptability. We need models for adjudications which we can combine with models for multi-version software to produce reliability models for complete fault-tolerant systems. We need understanding of responses to alarm indications, i.e. what happens when the adjudicator cannot make a decision?

Understanding of fault-tolerant structures

These include, but should not be restricted to, N-version programming and recovery blocks. Different kinds and degrees of diversity may be available at different levels of nesting. This raises issues of optimality: what is the best architecture for a particular problem? Experimental studies should be conducted of the efficacy of different structures in delivering reliability.

Understanding of factors affecting the efficacy of fault-tolerance

To what extent does this depend on the problem? For example, is there less to be gained from fault-tolerance for a complex problem? Are there other identifiable factors influencing success? Are there metrics for this which would allow us to predict the extent of likely success? To what extent does success depend on the desired reliability level, for example is there a law of diminishing returns for ultra-high reliability? What is the impact of the adjudicator: does success depend on the complexity (and hence likely unreliability) of the adjudicator?

Understanding the problem of ultra-dependability

It is particularly important to distinguish here between achievement and an assurance that a particular level of dependability has been achieved in a particular context. Since the issue of ultra-dependability is often associated with contexts in which failures have catastrophic consequences, it is important to measure what has actually been achieved. Confidence in the fault-tolerant methodology, and even evidence of its past successes, will be no substitute for confidence in the dependability of a particular fault-tolerant system. Can we **measure** this level of dependability? Can we **achieve** it? Does achievement of ultra-dependability require special techniques (what are they?) or merely the extensive use of ordinary fault-tolerance?

Dependability modelling

We need reliability models which combine version information, probably obtained from reliability growth modelling, with information about the particular fault-tolerant structure. These models could be particularly useful if software reusability were to fulfil its promise. Reusable modules could be expected to become available with very long failure histories, and therefore have very high assured reliabilities.

2.5 Specification Issues

It became apparent early in our discussions that techniques for the achievement of software reliability through fault-tolerance depend critically on the accuracy of the specification.

We decided to widen our brief to include recommendations for research here:

Investigation of diverse specifications

Use of a specification as the basis for a simulation model of the environment

The idea here is that the output of the model could be checked against the real environment, thus checking the specification.

Understanding of the qualities of a specification that engender misinterpretation by implementers

Issues concerning detailed (tight) specification versus global (loose) specification, particularly as they relate to diversity

Use of rapid prototyping to refine high-level specifications

Specification of dependability objectives themselves

2.5 Other Issues

The following issues do not easily fall into earlier headings, and some of them are relevant to several areas. There is no suggestion that these are less important.

Data issues

Many of the unresolved questions depend on observing the behaviour of actual systems. Current experimentation inevitably tends to take place in relatively unrealistic contexts. These interesting results need validation in the real world. Data should be collected from real systems, under test in simulators and in operation. The latter is particularly important and should in certain cases be enforced by licensing agencies. It is important that data should be as complete as is practicable, for example, the occasions on which the user is invoked in operational fault-tolerant systems should be recorded.

Management and quality control

If, using a hardware analogy, we partition the production of software into theory, development, manufacturing and maintenance, can we use standard quality control and product assurance techniques? We need a life-cycle model of fault-tolerant software. What are the organisational implications? Can we currently agree on a suite of statistical tools/models for use during single version development for tracking of reliability achievement, deciding when to stop testing, etc? At what point in the development process should we bind hardware to software to formulate a system model?

Load dependence

We need to monitor real systems to gain understanding of the extent to which reliability depends on load. Can we model the dependence so that the reliability for a particular load can be predicted? Alternatively, can we design systems in which such dependence is absent?

Modelling of failure criticality with respect to delivered services

Most of the questions which have been posed earlier about reliability can also be asked about the **criticality** of failures. For example: how does it depend on the specification, on the development process, on the use of multiple error detection mechanisms, on forced versus inherent diversity, on hardware (transient) faults, etc? We need metrics and models so that we can extend our reliability statements, which essentially are only about **frequencies** of failures, to include predictions about the consequential **costs** of failures.

Comparative evaluation of different techniques for dependability achievement

System dependability can be achieved by the use of several techniques, either singly or in combination, for example: design fault-tolerance, formal verification, testing. System developers need information on the comparative efficacy of these approaches so that proper cost-benefit trade-offs can be made.

Detection of faults by type

Can we steer testing towards particular types of faults, for example, those with high occurrence rates, those with severe consequences? Can we do the same for adjudicators? Can we get near 100% protection against special types of faults, and know that we have done this in a particular context? Are the faults which remain, after our current fault-removal techniques have been applied, different in kind from the ones we successfully remove? If so, in what way?

Supervisory environments

Very dependable environments are essential for the implementation of diverse computing, both hardware and software.

Relationship between verification and dependability measurement

Formal verification is often thought of as an all-or-nothing exercise. In practice, for all except very small programs, "verification" stops short of a complete proof of correctness. Can we get measures of such partial verification and model their relationship to dependability?

3. Conclusion

In this short chapter we have not attempted to provide a summary of the intense and wide-ranging discussion which took place during what we hope and believe to have been a stimulating and productive three days. Instead, we have concentrated on recording the output from this working meeting in terms of suggested avenues for further research and investigation. The theory and practice of software fault tolerance can hardly yet be considered as well established; if the research proposals presented here are followed up then our ability to model, and thereby measure and predict, the effect of diversity and adjudication strategies on software reliability should certainly improve - in which case the efforts of the workshop participants to disentangle the many relevant factors will have been worthwhile.

Acknowledgements

The success of the workshop owed everything to our participants. These were:

Prof. Tom Anderson, Dr. Dorothy M. Andrews, Dr. Jean Arlat, Prof. Al Avizienis, Mr. Mel Barnes, Dr. William C. Carter, Dr. Alain Costes, Ms. Janet R. Dunham, Mr. Mike Dyer, Mr. Dave Eckhardt, Dr. Jim W. Gault, Mr. Jack Goldberg, Mr. Gunnar Hagelin, Prof. Ravi Iyer, Prof. John P.J. Kelly, Prof. John C. Knight, Prof. Hermann Kopetz, Dr. Jean-Claude Laprie, Prof. Nancy G. Leveson, Prof. Bev Littlewood, Prof. David F. McAllister, Dr. John McHugh, Prof. John F. Meyer, Mr. Earle Migneault, Prof. Doug Miller, Mr. Marco Mulazzani, Ms. Phyllis M. Nagel, Prof. Brian Randell, Dr. Francesca Saglietti, Dr. Lorenzo Strigini, Dr. Pascal J. Traverse, Mr. Udo Voges, Mr. Larry J. Yount.

The organising committee was Tom Anderson, Jim Gault, Jean-Claude Laprie, Bev Littlewood and Earle Migneault.

The workshop was funded by the US Army European Research Office. Grace Palmer was responsible for administration and organisation.

7

Conclusion

The papers which are included in this volume as well as additional reports presented at the Workshops in Baden and in Badgastein demonstrate that the use of Software Diversity in systems with high dependability requirements is a solution for achieving the anticipated goals.

We have seen that Software Diversity has many faces. It can be applied in two and in more than two versions, on a small scale (module level) or on a large scale (system level), as static (N-version programming) or dynamic redundancy (recovery blocks), for development purposes (testing only) or use in the final application.

The most often used diversity aspects were one or more of the following:
- independent teams
- different languages
- different algorithms
- different environment.

Sometimes, software diversity is joint by hardware diversity. Hardware design errors are nowadays considered as possible as software design errors, raising complexity and integration being one reason.

The use of fault avoidance techniques alone is not sufficient to guarantee the required level of reliability and safety in high risk applications. Fault tolerance techniques have to be used in addition.

Which of the fault avoidance techniques and which of the fault tolerance techniques should be used together is a question which can not be answered in general. Application specific judgement is necessary. Software Diversity is one of the competing fault tolerance techniques. Its merits as well as its drawbacks have to be evaluated before coming to a conclusion.

This evaluation is not trivial. Many open questions remain, and additional research and experimentation is necessary to solve them. This includes:

- What are the effects of diversity on the system, e. g. related to development, management, cost, maintenance, execution?
- How can diversity between solutions be measured?
- In which phases and to what extend should diversity be applied, where are the highest benefits?
- Which fault avoidance techniques are useful and necessary together with software diversity?
- Which testing techniques are the best partners to software diversity?
- How can fault tolerance techniques, software diversity in particular, best be modelled?

Despite these open questions, the application of software diversity in real life should not be postponed. Sufficient knowledge is present to guarantee an increase in dependability if this technique is applied correctly. And furthermore, real life use can also assist in getting answers to the above raised questions.

We hope that in some years from now a new volume on Software Diversity can report on success and solutions in some of the mentioned areas.

Appendix

IFIP WG 10.4 "Reliable Computing and Fault Tolerance"

Workshop on Design Diversity in Action

27-28 June 1986 in Baden/Vienna(Austria)

Program

Session 1: University Experiments

  T. Anderson, Univ. of Newcastle upon Tyne (GB)
  Recovery Block Experiment

  J. Kelly, UCLA; J. Knight, UVA (USA)
  NASA NCSU/UCLA/UIUC/UVA Experiment

Session 2: Nuclear Applications

U. Voges, Kernforschungszentrum Karlsruhe (D)
Use of Software Diversity in Experimental Reactor Safety Systems

P. Bishop, CERL (GB)
PODS - Project on Diverse Software

U. Voges, KfK (D)
EPRI - Experiment on Software Diversity

Session 3: Railway Applications

G. Hagelin, LM Ericsson (S)
Gothenburg Interlocking System

G. Hagelin, LM Ericsson (S)
Automatic Train Control System

Session 4: Flight Applications

P. Traverse, Aerospatiale (F)
Airbus and ATR System Architecture and Specification

N. Wright, GEC Avionics (GB)
Airbus 310 Implementation Issues

L. Yount, Sperry Corporation (USA)
Use of Diversity in Boeing Airplanes

T. Anderson, U. of Newcastle (GB)
Other Flight Applications

Session 5: General Discussion

Ad-hoc Brief Contributions from Participants

Discussion on 'Merits and Future of Design Diversity'

8

Annotated Bibliography

Annotated Bibliography on Software Diversity

Udo Voges
Kernforschungszentrum Karlsruhe GmbH
Institut für Datenverarbeitung in der Technik
Postfach 3640, D-7500 Karlsruhe 1
Federal Republic of Germany

This annotated bibliography is an attempt to list all relevant material which is related to software diversity. Nevertheless it will be incomplete and biased by the author. The author would welcome any further information on publications and work which should be contained in such a bibliography for later editions.

1. J. M. Adams, "On the Practicality of Software Redundancy," in *Proc. 20th Hawaii Intern. Conf. on System Sciences*, Vol. 2, pp. 31-40, Kailua-Kona, HI, USA, 6-9 January 1987.

    Software diversity is achieved through procedural and nonprocedural versions of the same program, with consistency checks between the versions. The advantages and the problems with this approach are discussed.

2. P. E. Ammann and J. C. Knight, "Data Diversity: An Approach to Software Fault Tolerance," in *Proc. 17th Intern. Symp. on Fault-Tolerant Computing FTCS' 17*, pp. 122-126, Pittsburgh, PA, USA, 6-8 July 1987.

3. M. Ancona, A. Clematis, G. Dodero, E. B. Fernandez, and V. Gianuzzi, "A System Architecture for Software Fault Tolerance," in *Proc. 3rd Intern. Conf. Fault-Tolerant Computing Systems*, Vol. IFB 147, pp. 273-283, Bremerhaven, Germany, 9-11 September 1987.

4. T. Anderson and R. Kerr, "Recovery Blocks in Action: A System Supporting High Reliability," in *Proc. 2nd Intern. Conf. on Software Engineering*, pp. 447-457, San Francisco, CA, USA, 13-15 October 1976.

    A brief account is presented of the recovery block scheme, together with a description of a new implementation of the underlying cache mechanism. A prototype system has been constructed to test the viability of these techniques by executing programs containing recovery blocks on an emulator for the proposed architecture.

5. T. Anderson and P. A. Lee, *Fault Tolerance: Principles and Practice,* Prentice Hall, Englewood Cliffs, NJ, USA, 1981.

    This book is a basic introduction into the area of fault tolerance. It introduces different techniques, including the recovery block approach.

6. T. Anderson and J. C. Knight, "A Framework for Software Fault Tolerance in Real-Time Systems," *IEEE Trans. on Software Engineering*, Vol. SE-9, No. 3, pp. 355-364, May 1983.

    A classification scheme for errors and a technique for the provision of software fault tolerance in cyclic real-time systems is presented. This technique is useful for application with the recovery block technique.

7. T. Anderson, "Can Design Faults be Tolerated?," in *Proc. 2nd GI/NTG/GMR-Fachtagung Fehlertolerierende Rechensysteme*, Vol. IFB 84, pp. 426-433, Bonn, Germany, 19-21 September 1984.

8. T. Anderson, "Fault Tolerant Computing," in *Resilient Computing Systems*, Ed. T. Anderson, Collins, London, 1985.

9. T. Anderson, D. N. Halliwell, P. A. Barrett, and M. R. Moulding, "An Evaluation of Software Fault Tolerance in a Practical System," in *Proc. 15th Intern. Symp. on Fault-Tolerant Computing FTCS' 15*, pp. 140-145, Ann Arbor, MI, USA, 19-21 June 1985.

    Description of an experiment with recovery blocks which demonstrated the increase of reliability through the use of this technique. (Compare contribution in this book.)

10. T. Anderson, P. A. Barrett, D. N. Halliwell, and M. R. Moulding, "Software Fault Tolerance: An Evaluation," *IEEE Trans. on Software Engineering*, Vol. SE-11, No. 12, pp. 1502-1510, December 1985.

This paper describes an experiment at the University of Newcastle upon Tyne with recovery blocks as software fault tolerance technique. The problem was a naval command and control system, a real-time system. The results of the experiment show that a reliability improvement of about 75% could be achieved. (Compare contribution in this book.)

11. T. Anderson, "A Structured Decision Mechanism for Diverse Software," in *Proc. 5th Symposium on Reliability in Distributed Software and Database Systems*, pp. 125-129, Los Angeles, CA, USA, 13-15 January 1986.

Description of a decision mechanism which can be used as well for n-version programming as well as for recovery blocks. Different strategies for a filter and an arbiter are explained, having different application areas.

12. H. S. Andersson and G. Hagelin, "Computer Controlled Interlocking System," *Ericsson Review*, No. 2, pp. 74-80, 1981.

The interlocking system of LM Ericsson is described, which is installed in Gothenburg and Malmö, Sweden. It incorporates two independently developed programs, which run in the same computer, leading to a fail-safe action in case of differences. (Compare contribution by Hagelin in this book.)

13. J. Arlat, "Design of a Microcomputer Tolerating Faults Through Functional Diversity," Dr. Eng. dissertation (in French), National Polytechnic Institute, Toulouse, F, April 1979.

Design of a system with two diverse microprocessors, a monolithic microprocessor (TMS 9900) and its emulation at the instruction set level by a bit-slice microprocessor (AMD 2900 series) for detecting similar errors and tolerating externally induced transient faults.

14. A. Avižienis, "Fault-Tolerance and Fault-Intolerance: Complementary Approaches to Reliable Computing," in *Proc. Intern. Conf. on Reliable Software*, pp. 458-464, Los Angeles, CA, USA, 21-23 April 1975.

Fault tolerance and fault intolerance in the system design - hardware as well as software - are compared with each other. Fault tolerance can be achieved by hardware redundancy, software redundancy and time redundancy. The use of redundant programming with parallel or sequential execution and a comparison is proposed in analogy to the commonly known and applied hardware redundancy.

15. A. Avižienis, "Fault-Tolerant Computing - Progress, Problems, and Prospects," in *Proc. IFIP Information Processing 77*, pp. 405-420, Toronto, Canada, August 1977.

This paper presents an integrated view of three main aspects of fault tolerance: pathology of faults, implementation of tolerance, and analysis of fault tolerant designs. Several current obstacles to a wider acceptance of fault tolerance in system design are identified, and some directions for the advancement of the understanding and use of fault tolerance, including software diversity, are suggested.

16. A. Avižienis and L. Chen, "On the Implementation of N-Version Programming for Software Fault-Tolerance During Program Execution," in *Proc. COMPSAC'77*, pp. 149-155, Chicago, IL, USA, November 1977.

    A pilot experiment in N-version programming is described and an evolving methodology for this form of programming is outlined. 27 independent versions of a program were implemented, and evaluations on 3-version systems were made.

17. A. Avižienis, "Fault Tolerance: The Survival Attribute of Digital Systems," *Proc. IEEE*, Vol. 66, No. 10, pp. 1109-1125, October 1978.

18. A. Avižienis, "Design Diversity - The Challenge for the Eighties," in *Proc. 12th Intern. Symp. on Fault-Tolerant Computing FTCS'12*, pp. 44-45, Santa Monica, CA, USA, June 1982.

19. A. Avižienis, "Design Diversity: An Approach to Fault Tolerance of Design Faults," in *AFIPS Vol. 53*, 1984.

20. A. Avižienis and J. P. J. Kelly, "Fault-Tolerance by Design Diversity: Concepts and Experiments," *IEEE Computer*, Vol. 17, No. 8, pp. 67-80, August 1984.

21. A. Avižienis, P. Gunningberg, J. P. J. Kelly, R. T. Lyu, L. Strigini, P. J. Traverse, K. S. Tso, and U. Voges, "Software Fault-Tolerance by Design Diversity; DEDIX: A Tool for Experiments," in *Proc. IFAC Workshop SAFECOMP'85*, pp. 173-178, Como, Italy, 1-3 October 1985.

    (Compare contribution by Avižienis et al in this book.)

22. A. Avižienis, P. Gunningberg, J. P. J. Kelly, L. Strigini, P. J. Traverse, K. S. Tso, and U. Voges, "The UCLA DEDIX System: A Distributed Testbed for Multiple-Version Software," in *Proc. 15th Intern. Symp. on Fault-Tolerant Computing FTCS'15*, pp. 126-134, Ann Arbor, MI, USA, 19-21 June 1985.

    (Compare contribution by Avižienis et al in this book.)

23. A. Avižienis, "The N-Version Approach to Fault-Tolerant Software," *IEEE Trans. Software Engineering*, Vol. SE-11, No. 12, pp. 1491-1501, December 1985.

    The reasons for N-version programming, its history at UCLA, and the results achieved are explained. The DEDIX system, a distributed supervisor and testbed for N-version software, is described (50 ref.). (Compare contribution by Avižienis et al in this book.)

24. A. Avižienis and J.-C. Laprie, "Dependable Computing: From Concepts to Design Diversity," *IEEE Proceedings*, Vol. 74, No. 5, pp. 629-638, May 1986.

    Description of a conceptual framework for expressing the attributes of what constitutes dependable and reliable computing and of the use of design diversity to cope with design faults in hardware and software.

25. A. Avižienis and D. A. Rennels, "The Evolution of Fault Tolerant Computing at the Jet Propulsion Laboratory and at UCLA: 1955 - 1986," in *The Evolution of Fault-Tolerant Computing*, Ed. A. Avižienis, H. Kopetz, and J.-C. Laprie, pp. 141-191, Springer-Verlag Wien New York, 1987.

Includes the history of N-version programming at UCLA.

26. B. Bhargava, "Software Reliability in Real-Time Systems," in *Proc. National Computer Conference*, pp. 297-309, Chicago, 1981.

27. P. Bishop, D. Esp, M. Barnes, P. Humphreys, G. Dahll, J. Lahti, and S. Yoshimura, "Project on Diverse Software - An Experiment in Software Reliability," in *Proc. IFAC Workshop SAFECOMP'85*, pp. 153-158, Como, Italy, 1-3 October 1985.

An international experiment which makes use of software diversity is described. (Compare contribution in this book.)

28. J. P. Black, D. J. Taylor, and D. E. Morgan, "A Case Study in Fault Tolerant Software," *Software - Practice and Experience*, Vol. 11, pp. 145-157, 1981.

29. S. S. Brilliant, J. C. Knight, and N. G. Leveson, "Analysis of Faults in an N-Version Software Experiment," TR-86-20, University of Virginia, September 1986.

30. S. S. Brilliant, J. C. Knight, and N. G. Leveson, "The Consistent Comparison Problem in N-Version Software," *ACM Sigsoft SEN*, Vol. 12, No. 1, pp. 29-34, January 1987.

The problem of comparison of the results in an N-version system is explained, which exists not only due to software errors, but also with correct versions due to rounding errors, algorithmic differences in the versions etc. Some possible solutions are presented, but no general applicable solution can be given (10 ref.).

31. J. E. Brunelle and D. E. Eckhardt, "Fault-Tolerant Software: Experiment with the SIFT Operating System," in *Proc. AIAA/ACM/NASA/IEEE Computers in Aerospace V Conf.*, pp. 355-360, Long Beach, CA, USA, 21-23 October 1985.

The N-version programming and recovery block techniques were implemented on a portion of the SIFT operating system. The results indicate that, to effectively implement fault-tolerant software design techniques, system requirements will be impacted and suggest that retrofitting fault-tolerant software on existing designs will be inefficient and may require system modification.

32. A. K. Caglayan and D. E. Eckhardt, "Systems Approach to Software Fault Tolerance," in *Proc. AIAA/ACM/NASA/IEEE Computers in Aerospace V Conference*, pp. 361-369, Long Beach, CA, USA, 21-23 October 1985.

The issues involved in applying fault-tolerant techniques to flight software are discussed. The problem of software and system instability and the effect of fault-tolerant software on it are explained. The use of recovery blocks and of N-version programming is analyzed (16 ref.).

33. S. D. Cha, "A Recovery Block Model and its Analysis," in *Proc. IFAC Workshop Safety of Computer Control Systems 1986 (SAFECOMP'86)*, pp. 21-26, Sarlat, France, 14-17 October 1986.

34. A. Cheilan and J.-C. Laprie, "Software Fault Tolerance: Why, How and How Much," LAAS Report No. 87077, March 1987.

This paper is devoted to software fault tolerance by means of design diversity. First, a rational for software fault tolerance for critical computing systems is presented.

Then, a unified presentation of the methods for software fault tolerance is given. Finally, an analysis of software fault tolerance concerning cost and reliability is discussed.

35. L. Chen and A. Avižienis, "N-Version Programming: A Fault-Tolerance Approach to Reliability of Software Operation," in *Proc. 8th Intern. Symp. on Fault-Tolerant Computing FTCS'8*, pp. 3-9, Toulouse, France, 21-23 June 1978.

   Introduction to N-version programming and the mechanisms for execution: comparison vector, comparison status indicator, and synchronization mechanism. A possible implementation in PL/1 is given. The problem of inexact voting is discussed. The experiment of N-version programming of a mini text editing system is described.

36. L. Chen, "Improving Software Reliability by N-Version Programming," ENG-7843, UCLA, Computer Science Department, Los Angeles, CA, USA, August 1978.

   This thesis introduces the concepts of comparison vector (c-vector), comparison status indicator (cs-vector), cross-check points (cc-points) and the comparison algorithm, including inexact voting. It is the conceptual basis for the DEDIX system (compare Avižienis et al 1985).

37. D. G. Clews, "Post Certification Aspects of Digital Systems - Pain or Pleasure for the Operator?," in *Proc. Conf. Intern. Federation of Airworthiness*, Singapore, June 1983.

38. J. P. Considine and J. J. Myers, "MARC: MVS Archival Storage and Recovery Program," *IBM Systems Journal*, Vol. 4, pp. 378-397, 1977.

39. J. D. Cummins, "Fault Detection Using Inverse Transfer Characteristic Software," in *Proc. IFAC Workshop SAFECOMP'86*, pp. 73-81, Sarlat, France, 14-17 October 1986.

   Some general ideas on the theme of the inverse system transfer characteristic for fault detection are investigated. Schemes are described for providing diverse versions of software for fault detection leading to improved fault tolerance for highly reliable safety system software.

40. G. Dahll, U. S. Jorgensen, J. M. Hölsö, and J. Lahti, "Examination of Methods for Production and Testing of Highly Reliable Programmes," in *Enlarged Halden Programme Group Meeting*, Fredrikstad, Norway, 6-9 June 1977.

41. G. Dahll and J. Lahti, "An Investigation of Methods for Production and Verification of Highly Reliable Software," in *Proc. IFAC Workshop SAFECOMP'79*, pp. 89-94, Stuttgart, Germany, 16-18 May 1979.

   Description of one of the early experiments with software diversity. As an example a part of a reactor safety system was implemented by two independent teams in two different languages.

42. P. A. Davies, "The Latest Developments in Automatic Train Control," in *Proc. Intern. Conf. on Railway Safety Control and Automation Towards the 21st Century*, pp. 272-279, London, United Kingdom, 25-27 September 1984.

For an automatic train control system to be used in Singapore Mass Rapid Transit Railway, part of the system is realized with two diverse redundant microprocessors having different design objectives and diverse software in order to minimize common mode failures. The design of the total automatic train control system is explained.

43. G. Demars, E. Girard, and J.-C. Rault, "APL in a Two-Step Programming Technique for Developing Complex Programs," in *Proc. APL Congress 73*, pp. 83-90, Copenhagen, Denmark, 22-24 August 1973.

A program is first developed in APL and then coded in Fortran. The two versions are executed and the results from testing are compared for error detection. This is an application of software diversity (dual coding) for testing purposes only.

44. K. H. Dorato, "Fault Tolerant Multi-Version Software: The Problem of Similar Errors," Master Thesis, University of California Los Angeles, 1986.

This thesis is a first evaluation of the 20 versions produced in the four University experiment described in /Kelly et al 1986/. It is based on the analysis of 1.000 test runs and gives several figures for correct and erroneous versions, single as well as TMR-systems. Basic result is a gain in correctness, and a not neglectable number of similar errors.

45. J. R. Dunham, "Software Errors in Experimental Systems Having Ultra-Reliability Requirements," in *Proc. 16th Intern. Symp. on Fault-Tolerant Computing FTCS' 16*, pp. 158-164, Wien, Austria, 1-4 July 1986.

46. J. R. Dunham and L. A. Lauterbach, "Reliability Analysis of a Three-Version Software System," in *Proc. COMPSAC' 86*, pp. 484-490, Chicago, IL, USA, 8-10 October 1986.

A discussion is presented of the reliability analysis of an experimental three-version software system configured with three alternate decision rules.

47. J. R. Dunham, "Experiments in Software Reliability: Life-Critical Applications," *IEEE Trans. on Software Engineering*, Vol. SE-12, No. 1, pp. 110-123, January 1986.

48. W. R. Dunn, "Software Reliability: Measures and Effects in Flight Critical Digital Avionics Systems," in *Proc. IEEE/AIAA 7th Digital Avionics Systems Conf.*, pp. 664-669, Fort Worth, TX, USA, 13-16 October 1986.

Software reliability and related models are discussed. The use of N-version programming and recovery blocks in flight critical applications for software reliability improvement is mentioned. The problem of remaining common-mode errors is raised.

49. D. E. Eckhardt and L. D. Lee, "An Analysis of the Effects of Coincident Errors on Multi-Version Software," in *Proc. AIAA/ACM/NASA/IEEE Computers in Aerospace V Conference*, pp. 370-373, Long Beach, CA, USA, 21-23 October 1985.

Presentation of the Eckhardt & Lee Model for the effects of coincident errors on N-version software systems. Some estimation models for the expected probability of system failure for an N-version system are given (5 ref.).

50. D. E. Eckhardt and L. D. Lee, "A Theoretical Basis for the Analysis of Multiversion Software Subject to Coincident Errors," *IEEE Trans. on Software Engineering*, Vol. SE-11, No. 12, pp. 1511-1517, December 1985.

Coincident errors are called those errors, which occur jointly in more than one version of a fault-tolerant N-version system. A probabilistic framework for evaluating the influence of coincident errors on a multiversion system is developed. Conditions are given under which the N-version approach is superior to a single version approach. Later papers cite the presented model as the Eckhardt & Lee Model (14 ref.).

51. W. Ehrenberger and M. Kersken, "Zuverlässigkeitseigenschaften diversitärer Programmsysteme (Reliability of Diverse Programs - in German)," in *Proc. Fachtagung Prozeßrechner 1981*, pp. 230-239, München, Germany, 10-11 March 1981.

A reliability model for diverse systems is presented.

52. W. Ehrenberger, "Safety, Availability, and Cost Questions about Diversity," in *Proc. IFAC Conference on Control in Transportation Systems*, pp. 261-267, Baden-Baden, Germany, April 1983.

53. W. R. Elmendorf, "Fault-Tolerant Programming," in *Proc. 2nd Intern. Symp. on Fault-Tolerant Computing FTCS'2*, pp. 79-83, Newton, MA, USA, 19-21 June 1972.

54. R. S. Fabry, "Dynamic Verification of Operating System Decisions," *Communications of the ACM*, Vol. 16, No. 11, pp. 659-668, November 1973.

Use of different algorithms and different hardware for dynamic verification of user access rights.

55. F. Fetsch, L. Gmeiner, and U. Voges, "Entwurf eines hochzuverlässigen redundanten Mikrorechnernetzes (Design of a High Reliable Redundant Microcomputer Network - in German)," in *Proc. GI - 11. Jahrestagung*, pp. 317-326, München, Germany, 20-23 October 1981.

An integrated approach to the design of a safety-related nuclear protection system is described. The use of software diversity as one means to achieve reliable software is discussed. (Compare contribution by Voges in this book.)

56. O. Firschein and M. S. Fischler, "Fault Tolerance Hardware and Software Techniques for Communications Multiprocessors," in *Proc. National Electronics Conf.*, Vol. 29, pp. 57-61, 1974.

57. M. A. Fischler and O. Firschein, "A Fault Tolerant Architecture for Real Time Control Applications," in *Proc. 1st Annual Symp. on Computer Architecture*, Florida, USA, December 1973.

58. M. A. Fischler, O. Firschein, and D. L. Drew, "Distinct Software: An Approach to Reliable Computing," in *Proc. Second USA-Japan Computer Conference*, pp. 573-579, 1975.

This paper is an early paper with a fairly complete description of the topics related to software diversity. Distinct software in the sense of software diversity is introduced. The different application areas are mentioned: for system checkout, as standby spare in dynamic redundancy, and as static redundancy. The relation of distinct software to design errors as well as main types of software distinctness are discussed. Results from a first experiment are reported.

59. R. Frullini and A. Lazzari, "Use of Microprocessor in Fail-Safe on Board Equipment," in *Proc. Intern. Conf. on Railway Safety Control and Automation Towards the 21st Century*, pp. 292-299, London, United Kingdom, 25-27 September 1984.

   As a fail-safe design criterion, diversity is used in Italian train control systems using microprocessors. Functionally diverse programs are running in sequence, and their results are compared for error detection (1 ref.).

60. J. R. Garman, "The 'Bug' Heard 'Round the World," *ACM Sigsoft SEN*, Vol. 6, No. 5, pp. 3-10, October 1981.

   Description of the bug in the software of the space shuttle which caused a delay of the start of the shuttle. It was a synchronization problem between the redundant computer system consisting of four computers and the fifth one, supporting the backup flight software, due to a forgotten synchronization at the very beginning.

61. J. Gayen, "Ein Beitrag zum Thema Diversität in Sicherungseinrichtungen spurgebundener Verkehrssysteme (A Contribution to the Topic Diversity in Safety Systems for Railway Traffic Systems - in German)," *Signal + Draht*, Vol. 75, No. 1/2, pp. 12-15, 1983.

62. W. Geiger, L. Gmeiner, H. Trauboth, and U. Voges, "Program Testing Techniques for Nuclear Reactor Protection Systems," *IEEE Computer*, Vol. 12, No. 8, pp. 10-18, August 1979.

   This paper elaborates on different techniques to be used to achieve a reliable system. One of the constructive techniques mentioned is software diversity.

63. D. P. Geller, "Coding in Two Languages Boots Program Reliability," *Electronic Design*, Vol. 31, No. 7, pp. 161-170, March 1983.

64. T. Gilb, "Parallel Programming," *Datamation*, Vol. 20, No. 10, pp. 160-161, October 1974.

   An overview of the dual coding technique is given.

65. T. Gilb, "Data Engineering," *Studentlitteratur*, Lund, 1976.

   This book contains a seven page description of the dual programming approach. It mentions thirteen applications of this technique in Europe, USA and Australia, but mainly based on private communication. No real quantified data which is usable for further investigation is included. No own experience is mentioned. The value of this technique for on line purposes as well as during development only or for maintenance purposes is explained.

66. T. Gilb, *Software Metrics,* Studentlitteratur, Lund, Sweden, 1976.

   The use of dual code as an indirect measuring technique for the reliability and correctness of the software is explained. Its applicability even in larger projects (e.

g. 20 KLOC) is demonstrated.

67. E. Girard and J.-C. Rault, "A Programming Technique for Software Reliability," *1st IEEE Symposium on Computer Software Reliability*, pp. 44-50, 1973.

A two step program writing technique is proposed: first a high-level program in APL, and then the final version in Fortran. Both versions are checked against each other, making use of statistical testing techniques. The APL-version is not used for final installation (41 ref.).

68. L. Gmeiner and U. Voges, "Software Diversity in Reactor Protection Systems: An Experiment," in *Proceedings IFAC Workshop Safety of Computer Control Systems*, pp. 75-79, Stuttgart, Germany, 16-18 May 1979.

(Compare contribution by Voges in this book.)

69. L. Gmeiner and U. Voges, "Experimentelle Untersuchungen zur Software-Diversität (Experimental Evaluation of Software Diversity - in German)," in *KFK-PDV 179*, pp. 126-139, Kernforschungszentrum Karlsruhe, December 1979.

The need for software diversity in order to achieve reliable software is discussed. An experiment with software diversity is described.

70. J. Goldberg, "SIFT: A Provable Fault-Tolerant Computer for Aircraft Flight Control," in *Proc. IFIP Congress Information Processing 80*, pp. 151-156, Tokyo, Japan, 6-9 October 1980.

71. B. A. Golovkin, "Multiversion Programming and its Application," *Autom. & Remote Control*, Vol. 47, No. 7, pp. 877-903, July 1986.

Survey on multiversion programming, including N-version programming and recovery blocks, its theory and some experiments and applications. A classification of fault-tolerant systems is given, based on different types of identical and not identical redundancy.

72. C. J. Goring, "A Practical Approach to Diversity and Redundancy," in *IEE Colloquium on 'Programmable Electronic Systems and Safety - HSE Guidelines'*, Vol. Digest No. 74, pp. 7/1-3, IEE, London, England, 9 June 1987.

Description of a TMR-system which uses hardware diversity and provides tools for software diversity. The system is used for the Space Shuttle re-entry vehicle.

73. T. Grams, "Diversitäre Programmierung: Kein Allheilmittel (Software Diversity: No Cure-All - in German)," *Informationstechnik*, Vol. 28, No. 4, pp. 196-203, 1986.

Software Diversity is no Cure-all. Errors which result from thinking habits have to be avoided in the early design stages.

74. T. Grams, "Biased Programming Faults - How to Overcome them?," in *Proc. 3rd Intern. Conf. Fault-Tolerant Computing Systems*, Vol. IFB 147, pp. 13-23, Bremerhaven, Germany, 9-11 September 1987.

75. S. T. Gregory and J. C. Knight, "A New Linguistic Approach to Backward Error Recovery," in *Proc. 15th Intern. Symp. on Fault-Tolerant Computing FTCS' 15*, pp. 404-409, Ann Arbor, MI, USA, 19-21 June 1985.

76. W. Grigulewitsch, K. Meffert, and G. Reuß, "Aufbau elektrischer Maschinensteuerungen mit diversitärer Redundanz (Design of Electrical Machine Control Systems with Diverse Redundancy - in German)," BIA - Report 5/86, 1986.

Machines, which require manual operation, like cutting machines and moulding machines, are equipped with safety devices to protect man. Nowadays, these safety devices are computerized, and they need to be redundant. This report contains findings and experience with diverse redundant systems, gained during the safety technical analysis of those systems.

77. A. Grnarov, J. Arlat, and A. Avižienis, "On the Performance of Software Fault-Tolerance Strategies," in *Proc. 10th Intern. Symp. on Fault-Tolerant Computing FTCS' 10*, pp. 251-253, Kyoto, Japan, 1-3 October 1980.

This paper presents a comparison of processing time and reliability performance for the recovery block scheme and the n-version programming scheme. The respective models are introduced (9 ref.).

78. K.-E. Großpietsch and U. Voges, "Methoden der Fehlerbehandlung (Methods for Error Handling - in German)," *Informatik-Spektrum*, Vol. 9, No. 2, pp. 95-109, April 1986.

Overview article on different fault tolerance techniques for hardware and for software, including diversity.

79. P. Gunningberg and B. Pehrson, "Specification and Verification of a Synchronization Protocol for Comparison of Results," in *Proc. 15th Intern. Symp. on Fault-Tolerant Computing FTCS' 15*, pp. 172-177, Ann Arbor, MI, USA, 19-21 June 1985.

Specification and validation of a protocol, which can be used to synchronize distributed redundant software for an exchange and comparison of results. This protocol was implemented and used in the DEDIX system. (Compare contribution by Avižienis et al in this book.)

80. J. P. Hack, "Digitale Elektronik in Verkehrsflugzeugen (Digital Electronic in Airplanes - in German)," in *DGLR-Symposium*, Köln, Germany, 25-26 October 1983.

Description of a system for the A 310 containing two independent channels with the same function, which were programmed by two independent teams using two different languages.

81. H. Hecht, "Fault-Tolerant Software for Real-Time Applications," *ACM Computing Surveys*, Vol. 8, No. 4, pp. 391-407, December 1976.

The recovery block technique and its possible application are described. An associated reliability model is presented.

82. H. Hecht, "Fault-Tolerant Software," *IEEE Trans. Reliability*, Vol. R-28, No. 3, pp. 227-232, August 1979.

Includes a discussion of an elementary state transition model for recovery blocks.

83. H. Hecht, "Current Issues in Fault Tolerant Software," in *Proc. COMPSAC'80*, pp. 603-607, Chicago, IL, USA, 1980.

84. H. Hecht and M. Hecht, "Fault-Tolerant Software," in *Fault-tolerant computing. Theory and techniques. Vol. II*, Ed. D. K. Pradhan, pp. 658-696, Prentice-Hall, Englewood Cliffs, NJ, USA, 1986.

The motivation for fault-tolerant software is given. The design of fault-tolerant software using N-version programming and recovery blocks is discussed, and reliability models are presented.

85. G. Heiner, "Introduction to Software Reliability - A Key Issue of Computing Systems Reliability," in *AGARD-CP-261*, pp. 30.1-30.13, April 1979.

86. A. D. Hills, "A 310 Slat and Flap Control System Management and Experience," in *Proc. 5th DASC*, November 1983.

87. A. D. Hills, "Digital Fly-by-Wire Experience," in *Nato AGARD Conf.*, Edmunds AFB, CA, USA, October 1985.

88. E. F. Hitt and J. J. Webb, "A Fault-Tolerant Software Strategy for Digital Systems," *AIAA/IEEE 6th Digital Avionics Systems Conference*, pp. 211-216, Baltimore, MD, USA, 3-6 December 1984.

Fault-tolerant software techniques are discussed, like N-version programming and recovery block. The sources of software faults are described. An assessment of the current state of the art of fault-tolerant software is given, and the necessary research topics are discussed (25 ref.).

89. H. Hofer, "Erfahrungen mit Flight Standard Software (Experience with Flight Standard Software - in German)," in *Proc. DGLR-Symposium*, Köln, Germany, 25-26 October 1983.

Development of the A 310 Slat/Flap control system, starting with the system requirements specification, and then having two independent teams using different languages and different processors (Intel 8080 and Motorola 6800).

90. J. J. Horning, H. C. Lauer, P. M. Melliar-Smith, and B. Randell, "A Program Structure for Error Detection and Recovery," in *Proc. Intern. Symp. on Operating Systems*, pp. 171-187, Rocquencourt, France, 23-25 April 1974.

The recovery block concept is described as a method for structuring programs, including error detection and recovery facilities. The recursive cache is introduced as a mechanism for automatic back-tracking (7 ref.).

91. H. Hölscher and J. Rader, *Mikrocomputer in der Sicherheitstechnik (Microcomputers in Safety-Related Applications - in German)*, TÜV Rheinland, 1984.

This book gives a guideline on how to design and develop safety-related systems with microcomputers. A hierarchy of safety classes is introduced, and associated requirements for the design and the development are given. The use of diversity - hardware as well as software - is required in applications with the highest safety demands. 43 fault avoidance and fault tolerance techniques are briefly described (55 ref.).

92. R. K. Iyer, K. Ravishankar, and P. Velardi, "A Statistical Study of Hardware Related Software Errors in MVS," No. 83-12, Stanford University, Center for Reliable Computing, Stanford, CA, USA, October 1983.

About 11% of all software errors and over 40% of all software failures were found to be hardware related.

93. N. Jack, "Analysis of a Repairable 2-Unit Parallel Redundant System with Dependent Failures," *IEEE Trans. on Rel.*, Vol. R-35, pp. 444-446, October 1986.

Possible use to diverse vs. non-diverse systems.

94. E. J. Joyce, "The Art of Space Software," *Datamation*, Vol. 31, No. 22, pp. 30-34, 15 November 1985.

95. H. Kameda, "The Module Standby Organization: A Scheme for more Reliable Operating Systems," in *Proc. IEEE 3rd Texas Conf. on Computing Systems*, pp. 10-2/1 - 10-2/4, Austin, TX, USA, 1974.

The use of the recovery block technique in operating systems to increase the reliability. A monitor controls the correct execution of the primary modules and transfers control to the redundant modules in case of detected errors.

96. U. M. Kammerer, "Einsatzbedingungen für Mini- und Mikrorechner in Kernkraftwerken (Requirements for the Use of Mini- and Microcomputers in Nuclear Power Stations - in German)," *RWTÜV-Schriftenreihe*, Vol. 22, pp. 46-51, 1983.

97. K.-H. Kapp, R. Daum, E. Sartori, and R. Harms, "Sicherheit durch vollständige Diversität (Safety Through Complete Diversity - in German)," in *Proc. Fachtagung Prozeßrechner 1981*, pp. 216-229, München, Germany, 10-11 March 1981.

Description of theoretical concepts and a design for a train control system using hardware as well as software diversity. This was only an experimental system. Concurrent Pascal and Modula 1 were used as implementation languages.

98. K.-H. Kapp, "Eine Methode zur Konstruktion und Überprüfung sicherheitsrelevanter Automatisierungssoftware (A Method for the Construction and Validation of Safety-Relevant Automation Software - in German)," Diss., Universität Karlsruhe, Fakultät für Informatik, July 1985.

After a general introduction to the topic of fault tolerance, the design of a part of a remote control system, a safe picture generator system, is presented. It incorporates diversity by the use of different hardware (DEC LSI 11/23 and Z 8001), different languages (Modula I and seq./concurrent Pascal) and forced diversity with different data structures. The system did not go into operation.

99. J. P. J. Kelly, "Specification of Fault-Tolerant Multi-Version Software: Experimental Studies of a Design Diversity Approach," CSD-820927, UCLA, Computer Science Department, Los Angeles, CA, USA, September 1982.

100. J. P. J. Kelly and A. Avižienis, "A Specification Oriented Multi-Version Software Experiment," in *Proc. 13th Intern. Symp. on Fault-Tolerant Computing FTCS'13*, pp. 121-126, Milan, Italy, June 1983.

101. J. P. J. Kelly, A. Avižienis, B. T. Ulery, B. J. Swain, R.-T. Lyu, A. Tai, and K.-S. Tso, "Multi-Version Software Development," in *Proc. IFAC Workshop Safety of Computer Control Systems 1986 (SAFECOMP'86)*, pp. 43-49, Sarlat, France, 14-17 October 1986.

The paper gives a preliminary report on the four-university experiment sponsored by NASA in 1984-1986. The experiment consisted of programming 20 versions of a redundant strapped down inertial measurement unit. The aims of the experiment are described, and some basic findings are mentioned, but no data are given.

102. M. Kersken and W. Ehrenberger, "A Statistical Assessment of Reliability Features of Diverse Programs," *Reliability Engineering*, Vol. 2, pp. 233-240, 1981.

103. K. H. Kim, "Distributed Execution of Recovery Blocks: Approach to Uniform Treatment of Hardware and Software Faults," in *Proc. 4th Intern. Conf. Distributed Computing Systems*, pp. 526-532, San Francisco, CA, USA, 14-18 May 1984.

104. J. C. Knight and N. G. Leveson, "Correlated Failures in Multi-Version Software," in *Proc. IFAC SAFECOMP'85*, pp. 159-165, Como, Italy, 1-3 October 1985.

105. J. C. Knight, N. G. Leveson, and L. D. St. Jean, "A Large Scale Experiment in N-Version Programming," in *Proc. 15th Intern. Symp. on Fault-Tolerant Computing FTCS'15*, pp. 135-139, Ann Arbor, MI, USA, 19-21 June 1985.

An experiment is described in which 27 versions were independently programmed and tested with 1 million test cases. The analysis showed that independent programming does not imply independent failure behavior of the versions.

106. J. C. Knight and N. G. Leveson, "An Empirical Study of Failure Probabilities in Multi-Version Software," in *Proc. 16th Intern. Symp. on Fault-Tolerant Computing FTCS'16*, pp. 165-170, Wien, Austria, 1-4 July 1986.

Second analysis of the UCI-UVA experiment on software diversity. A more detailed analysis of the different reliability and safety figures for single versions, for 2-version systems and for 3-version systems are given. Generally, an increase in the reliability is achieved the more systems are combined. (Compare Knight et al 1985.)

107. J. C. Knight, "Data Diversity - A New Approach to Fault-Tolerant Software," in *Proc. 11th Annual Software Engineering Workshop*, NASA Goddard Space Flight Center, 3 December 1986.

Introduction of a technique to investigate the characteristics of failure regions in the input space. Use of slight variations in the input space shall detect the size of failure regions.

108. J. C. Knight and N. G. Leveson, "An Experimental Evaluation of the Assumption of Independence in Multiversion Programming," *IEEE Trans. on Software Engineering*, Vol. SE-12, No. 1, pp. 96-109, January 1986.

(Compare Knight et al 1985.)

109. R. Konakovsky, "On a Diversified Parallel Microcomputer System," in *Proc. IFAC Workshop SAFECOMP'79*, pp. 81-88, Stuttgart, Germany, 16-18 May 1979.

Three design structures of a microcomputer system with two parallel diverse hardware units are described. The failure detection capabilities are discussed.

110. H. Kopetz, "Software Redundancy in Real Time Systems," *Proc. Information Processing 74*, pp. 182-186, Stockholm, Sweden, 5-10 August 1974.

Introduction of static redundancy (triple modular redundancy) and standby redundancy (processing module, audit module, and standby module) for coping with software errors. With a simple model, the reliability gain from single module to TMR to standby redundancy is given (17 ref.).

111. H. Krebs and U. Haspel, "Ein Verfahren zur Software-Verifikation (A Technique for Software Verification - in German)," *Regelungstechnische Praxis*, Vol. 26, pp. 73-78, 1984.

The object code of a program was disassembled and translated back to higher level representations until a level was achieved, which was equivalent to the requirement specification. These two independently generated documents - the original requirement specification and the back translated one - are compared. This approach was used for the licensing of a railway safety system in Germany (0 ref.).

112. H. Krebs, "Verification of Safety Related Programs for a Maglev System," in *Proc. 5th IFAC/IFIP/IFORS Conf. Control in Transportation Systems*, pp. 357-363, Vienna, Austria, 8-11 July 1986.

Use of diverse back development in the verification procedure for a railway safety system. The program code is translated into higher level representations step by step until the level of the problem specification is reached. Original problem specification and back-translated specification are then compared.

113. J.-C. Laprie, "Dependability Evaluation of Software Systems in Operation," *IEEE Trans. on Software Engineering*, Vol. SE-10, No. 6, pp. 701-714, November 1984.

114. J.-C. Laprie, "Dependable Computing and Fault Tolerance: Concepts and Terminology," in *Proc. 15th Intern. Symp. on Fault-Tolerant Computing FTCS'15*, pp. 2-11, Ann Arbor, MI, USA, 19-21 June 1985.

This paper provides a conceptual framework for expressing the attributes of what constitutes dependable and reliable computing: the impairments of dependability (faults, errors, and failures), the means for dependability (fault-avoidance, fault-tolerance, error-removal, and error-forecasting), and the measures of dependability (reliability, availability, maintainability, and safety). Emphasis is being put on the dependability impairments and on fault-tolerance.

115. J.-C. Laprie, J. Arlat, C. Beounes, C. Hourtolle, and K. Kanoun, "Software Fault Tolerance," LAAS 86.044 (in French), April 1986.

Preparatory work performed for the Hermes space shuttle.

116. J.-C. Laprie, "Dependability: A Unifying Concept for Reliable Computing and Fault Tolerance," LAAS Report 86.357, December 1986.

117. J.-C. Laprie, J. Arlat, C. Beounes, K. Kanoun, and C. Hourtolle, "Hardware- and Software-Fault Tolerance: Definition and Analysis of Architectural Solutions," in *Proc. 17th Intern. Symp. on Fault-Tolerant Computing FTCS' 17*, pp. 116-121, Pittsburgh, PA, USA, 6-8 July 1987.

Introduction of N Self-Checking Programming as a third alternative besides the methods of Recovery Block and N-Version Programming and comparison of these three approaches.

118. D. Lardner, "Babbage's Calculating Engine; From the Edinburgh Review, July, 1834, No. CXX," in *Charles Babbage and His Calculating Engines*, Ed. E. Morrison, Dover Publications, Inc. New York, 1961.

First discovered mention of diversity of computations.

119. R. Lauber, "Safe Software by Functional Diversity," EWICS TC 7 WP 37, 1975.

120. N. G. Leveson, "An Empirical Study of Error Detection Using Self-Test," in *Proc. 11th Annual Software Engineering Workshop*, NASA Goddard Space Flight Center, 3 December 1986.

A comparison between the error detection capability of the voter in n-version programming and of acceptance tests in the program - specification based as well as code based - is made on the basis of a preliminary analysis of an experiment.

121. N. G. Leveson, "Software Fault Tolerance in Safety-Critical Applications," in *Proc. 3rd Intern. Conf. Fault-Tolerant Computing Systems*, Vol. IFB 147, pp. 1-12, Bremerhaven, Germany, 9-11 September 1987.

122. K.-J. Lin, "Resilient Procedures - an Approach to Highly Available System," in *Proc. IEEE 1986 Intern. Conf. on Computer Languages*, pp. 98-106, Miami, FL, USA, 27-30 October 1986.

A resilient procedure is proposed that actively executes a single-version or multi-version procedure on several sites to provide high availability for a distributed computer system.

123. O. Berg von Linde, "Computers Can Now Perform Vital Functions Safely," *Railway Gazette International*, pp. 1004-1007, November 1979.

Description of the LM Ericsson approach to achieve safety in computerized railway safety systems in use in Sweden and Taiwan. The systems are explained, the reasons for using software diversity as well as the way software diversity is realized. The use of independent teams, inverted and reverted data and checkpoints are the main features of the design. The two resulting programs run in the same microprocessor. (Compare contribution by Hagelin in this book.)

124. L. Liotta and D. Sciuto, "Static and Dynamic Redundancy: Proposal and Evaluation of Two Constructs of Software Fault Tolerance," in *Proc. 11th EUROMICRO Symposium on Microprocessing and Microprogramming: Microcomputers, Usage and Design.*, pp. 463-473, Brussels, Belgium, 3-6 September 1985.

Within the context of Ada, the two constructs of software fault tolerance, recovery blocks and N-version programming, are explained. An error model is introduced and the error space of the two constructs is described. A comparison of the two constructs is given. The problems associated with the voting on similar results are explained (15 ref.).

125. B. Littlewood and D. R. Miller, "A Conceptual Model of Multi-Version Software," CSR Technical Report, December 1986.

Based on the Eckhardt & Lee Model, it is shown that a duality exists between input choice and program choice. The model is enhanced by incorporating the effects of the use of diverse methodologies. An optimal method for allocating diversity between versions can be obtained for certain 1-out-of-n systems.

126. B. Littlewood and D. R. Miller, "A Conceptual Model of Multi-Version Software," in *Proc. 17th Intern. Symp. on Fault-Tolerant Computing FTCS'17*, pp. 150-155, Pittsburgh, PA, USA, 6-8 July 1987.

127. B. Littlewood and D. R. Miller, "A Conceptual Model of the Effect of Diverse Methodologies on Coincident Failures in Multi-Version Software," in *Proc. 3rd Intern. Conf. Fault-Tolerant Computing Systems*, Vol. IFB 147, pp. 263-272, Bremerhaven, Germany, 9-11 September 1987.

128. A. B. Long, C. V. Ramamoorthy, S. F. Ho, H. H. So, H. L. Reeves, and E. A. Straker, "A Methodology for the Development and Validation of Critical Software for Nuclear Power Plants," in *Proc. COMPSAC'77*, pp. 620-626, Chicago, IL, USA, November 1977.

Description of a methodology to be used in an experiment, incorporating dual programming (software diversity). For results see So 1979, Ramamoorthy 1979, Ramamoorthy 1981, Saib 1982.

129. S. V. Makam, "Design Study of a Fault-Tolerant Computer System to Execute N-Version Software," CSD-821222, UCLA, Computer Science Department, Los Angeles, CA, USA, December 1982.

130. L. Mancini and G. Pappalardo, "The Join Algorithm: Ordering Messages in Replicated Systems," in *Proc. IFAC Workshop SAFECOMP'86*, pp. 51-55, Sarlat, France, 14-17 October 1986.

131. D. J. Martin, "Dissimilar Redundancy for Fly-by-Wire Secondary Flight Controls," in *Proc. Advanced Flight Controls Symposium*, Colorado Springs, CO, USA, 1981.

132. D. J. Martin, "Dissimilar Software in High Integrity Applications in Flight Controls," in *Proc. AGARD Symp. on Software for Avionics, CPP-330*, pp. 36.1-36.13, The Hague, The Netherlands, September 1982.

Description of the use of software diversity in airplanes.

133. G. E. Migneault, "The Cost of Software Fault Tolerance," in *Proc. AGARD Symposium on Software for Avionics, CPP-330*, pp. 37.1-37.8, The Hague, The Netherlands, 1982.

134. J. S. Miller, "On Software Quality," in *Proc. 2nd Intern. Symp. on Fault-Tolerant Computing FTCS'2*, pp. 84-88, Cambridge, MA, USA, 1972.

135. D. E. Morgan and D. J. Taylor, "A Survey of Methods of Achieving Reliable Software," *IEEE Computer*, Vol. 10, No. 2, pp. 44-53, February 1977.

Introduction

136. M. A. Morris, "An Approach to the Desing of Fault-Tolerant Software," MS Thesis, Cranfield Institute of Technology, September 1981.

A two-version software system is proposed, whose results are compared. If no agreement is achieved, a comparison with an estimate generated from previous results is performed, using the closest one - if within some boundaries - as correct one. This approach is only usable in continuous processes. It is called CRAFTS (Cranfield Algorithm for Fault-Tolerant Software) or also known as Foodtaster.

137. M. R. Moulding, "Techniques for Achieving Software Fault Tolerance," in *IEE Colloquium on 'High Integrity Systems - Theory and Practice'*, Vol. Digest No. 112, pp. 2/1-11, London, England, 5 November 1986.

The basic concept of fault tolerance is presented. A short description of the recovery block approach and the N-version programming approach is given. The two approaches are compared with each other. Overview paper (16 ref.).

138. M. Mulazzani, "Reliability versus Safety," in *Proc. IFAC Workshop SAFECOMP'85*, pp. 141-146, Como, Italy, 1-3 October 1985.

139. M. Mulazzani, "Reliability and Safety in Electronic Interlocking," in *Proc. 5th IFAC/IFIP/IFORS Conf. Control in Transportation Systems*, pp. 321-328, Vienna, Austria, 8-11 July 1986.

The impact of different methods on the reliability and safety of a railway system is examined. The hardware and software structure of several interlocking systems are analyzed. The principles applied in these interlocking systems include software diversity.

140. N. N., "N-Version Simulator Interface. User's Guide," RTI/43U-2094-12, Research Triangle Institute, October 1983.

A special purpose system for monitoring and controlling the execution of a simulator is explained. The simulator itself is an environment for executing a three-version implementation of a radar tracking problem.

141. N. N., "Guidance on the Safe Use of Programmable Electronic Systems: Part 2: Safety Integrity Assessment. Draft Document," Health and Safety Executive, Bootle, United Kingdom, 1984.

In this guideline, it is proposed to use as a figure for the ratio between the common mode failure rate and the independent failure rate a value between 0.03 and 0.3 for homogeneous redundancy and between 0.001 and 0.1 for diverse redundancy. A checklist on diverse software assessment is given. The guideline gives further information on the assessment of software and hardware for safety-related applications.

142. N. N., "Redundancy Management Software Requirements Specification for a Redundant Strapped Down Inertia Measurement Unit," Version 2.0, Charles River Analytics, Research Triangle Institute, Research Triangle Park, N.C., 30 May 1985.

Specification of the problem used in the four university experiment (compare Kelly et al 1986.).

143. N. N., "Automatic Software Generation and Validation for Nuclear Power Plant Status Monitoring," EPRI-NP-4784-SR, Palo Alto, 31 October 1986.

Use of diverse backward decomposition of logic routines to compare with the original specification in order to verify the correctness.

144. N. N., "Requirements for Software for Use with Digital Processors," Navel Engineering Standard NES 620, United Kingdom Ministry of Defence, October 1986.

For high integrity software this standard requires, among other techniques, the use of software diversity as a fault tolerance technique.

145. P. M. Nagel and J. A. Skrivan, "Software Reliability: Repetitive Run Experimentation and Modeling," NASA CR-165836, 1982.

146. H. G. Nix, "Sichere Mikroprozessorsysteme für Schutzaufgaben bei der Prozeßautomatisierung (Safe Microcomputer Systems for Safety Functions in Process Automation - in German)," *Automatisierungstechnische Praxis*, Vol. 28, No. 3, pp. 130-135, 1986.

147. D. Nordenfors and A. Sjöberg, "Computer-Controlled Electronic Interlocking System ERILOCK 850," *Ericsson Review*, No. 1, pp. 11-17, 1986.

The interlocking system and its installation in Hallsberg, Sweden, is described. The reasons for and the use of software diversity are explained shortly. (Compare contribution by Hagelin in this book.)

148. D. J. Panzl, "A Method for Evaluating Software Development Techniques," *Journal of Systems and Software*, Vol. 2, No. 2, pp. 233-240, June 1981.

149. S. Pfleger, "Structuring Concepts for Robust Applications," in *Proc. COMPSAC'86*, pp. 420-426, Chicago, IL, USA, 8-10 October 1986.

150. J. R. Popovic, D. C. Chan, D. B. Burjorjee, and B. K. Patterson, "Computer Control in Candu Plants," in *Symposium on Advanced Nuclear Services, CAN/CNS Intern. Nuclear Conference*, Toronto, Canada, 8-11 June 1986.

Use of hardware and software diversity in a reactor safety shutdown system.

151. C. V. Ramamoorthy, F. B. Bastani, J. M. Favaro, Y. R. Mok, C. W. Nam, and K. Suzuki, "A Systematic Approach to the Development and Validation of Critical Software for Nuclear Power Plants," in *Proc. 4th Intern. Conf. Software Engineering*, pp. 231-240, München, Germany, 17-19 September 1979.

A methodology is proposed for the development and validation of nuclear power plant safety system software. It includes the use of tools, formal techniques, and dual programming (software diversity). The application of this methodology in an experiment and its results are described. (Compare Ramamoorthy et al 1981, Saib 1982.)

152. C. V. Ramamoorthy, Y. R. Mok, F. B. Bastani, G. H. Chin, and K. Suzuki, "Application of a Methodology for the Development and Validation of Reliable Process Control Software," *IEEE Trans. on Software Engineering*, Vol. SE-7, No. 6, pp. 537-555, November 1981.

A formal specification development and validation methodology is proposed for achieving reliable software. The techniques are applied in a project, including the independent development of two sets of specifications, and continuing with independent design, implementation and testing. In a final phase, the two programs were executed with a test data generator and a dual program monitor system. Besides the two development teams, a third independent team was working on testing, review and comparison (30 ref.). (Compare Ramamoorthy et al 1979, Saib 1982.)

153. B. Randell, "System Structure for Software Fault Tolerance," *IEEE Trans. on Software Engineering*, Vol. SE-1, No. 2, pp. 220-232, June 1975.

Introduction of recovery block technique.

154. B. Randell, P. A. Lee, and P. C. Treleaven, "Reliability Issues in Computing System Design," *ACM Computing Surveys*, Vol. 10, No. 2, pp. 123-165, June 1978.

This paper surveys the various problems involved in achieving very high reliability computing systems. Topics covered include protective redundancy, use of atomic actions, error detection techniques, and error recovery techniques (50 ref.).

155. B. Randell, "Design Fault Tolerance," in *The Evolution of Fault-Tolerant Computing*, Ed. A. Avižienis, H. Kopetz, and J.-C. Laprie, pp. 251-270, Springer-Verlag Wien New York, 1987.

Includes the history of recovery blocks at the University of Newcastle upon Tyne.

156. J.-C. Rault, "Extension of Hardware Fault Detection Models to the Verification of Software," in *Program Test Methods*, Ed. W. C. Hetzel, pp. 255-262, Prentice-Hall, Inc., Englewood Cliffs, NJ, USA, 1973.

The use of dual coding as a testing method is proposed. A model of the final program is written in a higher level language and both versions are executed with the same test data for error detection.

157. J. C. Rouquet and P. J. Traverse, "Safe and Reliable Computing on Board the Airbus and ATR Aircraft," in *Proc. IFAC Workshop SAFECOMP'86*, pp. 93-97, Sarlat, France, 14-17 October 1986.

Description of the use of computers with safety and reliability requirements in the airplanes Airbus and ATR.

158. F. Saglietti and W. Ehrenberger, "Software Diversity - Some Considerations about its Benefits and its Limitations," in *Proc. IFAC Workshop Safety of Computer Control Systems 1986 (SAFECOMP'86)*, pp. 27-34, Sarlat, France, 14-17 October 1986.

159. F. Saglietti and M. Kersken, "Quantitative Assessment of Fault-Tolerant Software Architecture," in *Proc. 3rd Intern. Conf. Fault-Tolerant Computing Systems*, Vol. IFB 147, pp. 284-297, Bremerhaven, Germany, 9-11 September 1987.

160. S. H. Saib, "Validation of Real-Time Software for Nuclear Plant Safety Applications," EPRI NP-2646, November 1982.

Description of the EPRI-project which involved the use of diversity in an experimental design of a computerized system for a nuclear power plant. Dual teams independently developed and tested the software using a formal specification language, structured programming, static test tools, and an automated test bed. The entire process was formally monitored by a third independent verification and validation team. The effectiveness of the concept is evaluated. (See also Long 1977, So 1978, Ramamoorthy 1979, Ramamoorthy 1981)

161. E. Schmidt, "Eine lebenswichtige System-Steuerung, die nicht ausfallen darf (A Life-Critical Control System which May not Fail - in German)," *Minimicro Magazin*, Vol. 2, No. 9, pp. 74-77, September 1986.

162. W. Schwier, *Private communication*, 1987.

No use of software diversity within DB due to the associated problems with synchronization of the alternatives and the related costs.

163. R. K. Scott, "Data Domain Modeling of Fault-Tolerant Software Reliability," Ph. D. Dissertation, North Carolina State Univ., Raleigh, NC, USA, 1983.

164. R. K. Scott, J. W. Gault, and D. F. McAllister, "The Consensus Recovery Block," in *Proc. Total Systems Reliability Symposium*, pp. 74-85, 1983.

The Consensus Recovery Block is introduced. It is a hybrid combination of aspects from the Recovery Block and from the N-Version Programming. Reliability models for comparison of the three approaches are presented.

165. R. K. Scott, J. W. Gault, and D. F. McAllister, "Modeling Fault-Tolerant Software Reliability," in *Proc. 3rd Symp. on Reliability in Distributed Software and Database Systems*, pp. 15-27, Clearwater Beach, FL, USA, 17-19 October 1983.

This paper presents fault-tolerant software reliability models based on component reliabilities. Two methods for estimating component reliabilities and the associated variances are given along with an approach for calculating the system reliability estimate variance. The derived models are used as a basis for discussing trade-offs between recovery blocks and N-version programming.

166. R. K. Scott, J. W. Gault, D. F. McAllister, and J. Wiggs, "Investigation of Version Dependence in Fault-Tolerant Software ," in *Proc. Avionics Panel Spring 1984 Meeting on Design for Tactical Avionics Maintainability*, 1984.

167. R. K. Scott, J. W. Gault, D. F. McAllister, and J. Wiggs, "Experimental Validation of Six Fault-Tolerant Software Reliability Models," in *Proc. 14th Intern. Symp. on Fault-Tolerant Computing FTCS' 14*, pp. 102-107, Orlando, FL, USA, June 1984.

Recovery block, N-version programming, and consensus recovery block reliability models and their experimental application are discussed.

168. R. K. Scott, J. W. Gault, and D. F. McAllister, "Fault-Tolerant Software Reliability Modeling," *IEEE Trans. on Software Engineering*, Vol. SE-13, No. 5, pp. 582-592, May 1987.

Reliability models for recovery block, N-version programming, and consensus recovery block are presented. The models as well as experimental application of them show that the consensus recovery block has the highest reliability. In addition, a simple cost model is presented (15 ref.).

169. Kang G. Shin and Yann-Hang Lee, "Evaluation of Error Recovery Blocks Used for Cooperating Processes," *IEEE Trans. on Software Engineering*, Vol. SE-10, No. 6, pp. 692-700, November 1984.

170. S. K. Shrivastava and A. A. Akinpelu, "Fault-Tolerant Sequential Progamming Using Recovery Blocks," Computing Laboratory Technical Report 122, University of Newcastle upon Tyne, United Kingdom, March 1978.

171. S. K. Shrivastava, "Concurrent Pascal with Backward Error Recovery: Language Features and Examples," *Software - Practice and Experience*, Vol. 9, No. 12, pp. 1001-1020, December 1979.

The programming language Concurrent Pascal has been extended to include some language features that facilitate the writing of fault-tolerant software. It is possible now to write operating systems with a measure of fault tolerance, and to support fault-tolerant user programs. The paper describes these language features and illustrates their use with the help of a few working examples.

172. A. Sjöberg, "Automatic Train Control," *Ericsson Review*, No. 1, pp. 22-29, 1981.

Description of the automatic train control system JZG 700, which incorporates two independently developed programs running in the same processor, and whose results are compared. Difference in the results leads to a fail-safe action. (Compare contribution by Hagelin in this volume.)

173. J. R. Sklaroff, "Redundancy Management Techniques for Space Shuttle Computers," *IBM J. Res. Develop.*, Vol. 20, pp. 20-28, January 1976.

174. H. So, C. Nam, H. Reeves, T. Albert, E. Straker, S. Saib, and A. B. Long, "Experience with a Specification Language in the Dual Development of Safety System Software," in *Proc. IFAC Workshop SAFECOMP'79*, pp. 161-167, Stuttgart, Germany, 16-18 May 1979.

Description of the first stage of the EPRI project on the use of diversity in the development of safety related software. (See also Long et al 1978, Ramamoorthy et al 1979, Ramamoorthy et al 1981, Saib 1982)

175. M. D. Soneru, "A Methodology for the Design and Analysis of Fault-Tolerant Operating Systems," PhD Dissertation, Illinois Institute of Technology, Chicago, IL, USA, May 1981.

176. B. J. Sterner, "Computerised Interlocking System - a Multidimensional Structure in the Pursuit of Safety," *IMechE Railway Engineer International*, pp. 29-30, November/December 1978.

Description of computerised interlocking system using two independently developed programs within one computer. First real life use of software diversity in a running application system, which has been in operational use since 1977. (Compare contribution by Hagelin in this book.)

177. J. Stocker and J. Rauch, "CSTS: Ein Software-Testsystem für den Tornado-Autopiloten (CSTS : A Cross Software Test System for the Tornado Autopilot - in German)," in *Proc. DGLR-Symposium*, Köln, Germany, 25-26 October 1983.

Use of diversity for testing purposes only. Starting from the same functional specification, separate independent designs were made. One system was implemented in C on a PDP using floating point arithmetic, the other one in Assembly language on a AFDS using fixed point arithmetic. The latter one was the final system used in the Tornado autopilot.

178. L. Strigini and A. Avižienis, "Software Fault Tolerance and Design Diversity: Past Experience and Future Evolution," in *Proc. IFAC Workshop SAFECOMP'85*, pp. 167-172, Como, Italy, 1-3 October 1985.

179. B. J. Swain, "Group Branch Coverage Testing of Multi-Version Software," UCLA-CSD-860013, Los Angeles, CA, USA, December 1986.

Group branch coverage testing is a variant of branch coverage testing designed for testing multi-version software.

180. J. R. Taylor and U. Voges, "Use of Complementary Methods to Validate Safety Related Software Systems," in *Proc. IFAC 7th Triennial World Congress*, pp. 731-737, Helsinki, Finland, 12-16 June 1978.

Use of diversity as one means to achieve reliable software.

181. J. R. Taylor, "Redundant Programming in Europe," *ACM Sigsoft SEN*, Vol. 6, No. 1, pp. 1-2, January 1981.

Mentions different experiments with and use of software diversity within Europe.

182. N. Theuretzbacher, "Using AI-Methods to Improve Software Safety," in *Proc. IFAC Workshop SAFECOMP'86*, Sarlat, France, 14-17 October 1986.

The N-version programming technique is compared with the newly introduced safety bag technique. Within a railway interlocking system, this technique consists of a checking part - are all actions by the control software safe - and an active part - guaranteeing that safety actions are really taken. The safety bag is judged to result in higher safety than N-version programming. The implementation of the safety bag is described.

183. J. E. Tomayko, "NASA's Manned Spacecraft Computers," *Annals of the History of Computing*, Vol. 7, No. 1, pp. 7-18, January 1985.

184. H. Trauboth, "Zuverlässigkeit von DV-Systemen - Eine systemtechnische Aufgabe (Reliability in Process Control Systems: A Duty for System Technique - in German)," in *Proc. Architektur und Betrieb von Rechensystemen*, Vol. IFB 78, pp. 271-295, Karlsruhe, Germany, 26-28 March 1984.

185. R. Troy and C. Baluteau, "Assessment of Software Quality for the Airbus A310 Automatic Pilot," in *Proc. 15th Intern. Symp. on Fault-Tolerant Computing FTCS'15*, pp. 438-443, Ann Arbor, MI, USA, 19-21 June 1985.

186. K. S. Tso, A. Avižienis, and J. P. J. Kelly, "Error Recovery in Multi-Version Software," in *Proc. IFAC Workshop Safety of Computer Control Systems 1986 (SAFECOMP'86)*, pp. 35-41, Sarlat, France, 14-17 October 1986.

187. K. S. Tso and A. Avižienis, "Community Error Recovery in N-Version Software: A Design Study with Experimentation," in *Proc. 17th Intern. Symp. on Fault-Tolerant Computing FTCS' 17*, pp. 127-133, Pittsburgh, PA, USA, 6-8 July 1987.

Use of forward error recovery in a N-version system with a minority of failed versions.

188. K. S. Tso, "Error Recovery in Multi-Version Software," CSD-870013, UCLA, Los Angeles, CA, USA, March 1987.

The problem of error recovery in multi-version software systems is stated. A community error recovery algorithm is proposed as a solution. Its implementation in the UCLA DEDIX system and the results from an experiment with it are presented. (Compare contribution by Avižienis et al in this book.)

189. D. B. Turner, R. D. Burns, and H. Hecht, "Designing Micro-Based Systems for Fail-Safe Travel," *IEEE Spectrum*, Vol. 24, No. 2, pp. 58-63, February 1987.

The use of computerized systems in railroad, aircraft, and space vehicles is explained. Examples for application of homogeneous redundancy and diverse redundancy, hardware as well as software, are given.

190. P. Velardi, R. K. Iyer, and K. Ravishankar, "A Study of Software Failures and Recovery in the MVS Operating System," No. 83-7, Stanford University, Center for Reliable Computing, Stanford, CA, USA, July 1983.

191. P. Velardi and R. K. Iyer, "A Study of Software Failures in the MVS Operating System," *IEEE Transactions on Computers*, Vol. C-33, No. 6, pp. 564-568, June 1984.

192. U. Voges and J. R. Taylor, "A Survey of Methods for the Validation of Safety Related Software," in *Proc. IFAC Workshop SAFECOMP'79*, pp. 95-103, Stuttgart, Germany, 16-18 May 1979.

Besides other methods, the use of software diversity for achieving reliable systems is discussed.

193. U. Voges, F. Fetsch, and L. Gmeiner, "Use of Microprocessors in a Safety-Oriented Reactor Shut-Down System," in *Proc. EUROCON'82*, pp. 493-497, Lyngby, Denmark, 14-18 June 1982.

Description of a design for a computerized nuclear reactor safety system which makes use of triple modular redundancy and three diverse application software systems. (Compare contribution by Voges in this book.)

194. U. Voges, "Der Einsatz von Software-Diversität in Systemen mit hohen Zuverlässigkeitsanforderungen (The Use of Software Diversity in Systems with High Reliability Requirements - in German)," in *Proc. Software-Fehlertoleranz und -Zuverlässigkeit*, pp. 155-165, Bremerhaven, Germany, 1984.

195. U. Voges, "Application of a Fault-Tolerant Microprocessor-Based Core-Surveillance System in a German Fast Breeder Reactor," in *EPRI Seminar: Digital Control and Fault-Tolerant Computer Technology*, Scottsdale, AZ, USA, 9-12 April 1985.

The application of software diversity in the design of a reactor safety shut-down system is explained. Three independent program developments in three different languages were planned. (Compare contribution by Voges in this book.)

196. U. Voges, "Anwendung von Software-Diversität in rechnergesteuerten Systemen (The Application of Software Diversity to Computer Controlled Systems - in German)," *Automatisierungstechnische Praxis atp*, Vol. 28, No. 12, pp. 583-588, 1986.

The paper describes the state of the art of software diversity. It mainly reports on the IFIP-Workshop 'Design Diversity in Action', gives some basic definitions and states open problems.

197. G. Weber, L. Gmeiner, and U. Voges, "Methoden der Zuverlässigkeitsanalyse und -sicherung bei Hardware und Software (Methods for Reliability Analysis and Achievement for Hardware and Software - in German)," in *Proc. Zuverlässigkeit von Rechensystemen*, Ed. W. Görke, pp. 71-96, Karlsruhe, Germany, 28-29 September 1978.

Besides other methods, the use of diversity in hardware and software in order to achieve high reliable systems is discussed.

198. A. Y.-W. Wei, "Real-Time Programming with Fault Tolerance," PhD Dissertation, University of Illinois, Urbana, IL, USA, 1981.

199. H. O. Welch, "Distributed Recovery Block Performance in a Real-Time Control Loop," in *Proc. Real-Time Systems Symposium*, pp. 268-276, Arlington, VA, USA, 1983.

200. J. H. Wensley, "SIFT - Software Implemented Fault Tolerance," in *AFIPS Conf. Proc.*, Vol. 41, pp. 89-96, 1972.

201. J. F. Williams, L. J. Yount, and J. B. Flannigan, "Advanced Autopilot-Flight Director System Computer Architecture for Boeing 737-300 Aircraft," in *Proc. Fifth Digital Avionics Systems Conference*, Seattle, WA, USA, 30 October - 3 November 1983.

The evolution of flight control computers from analog to digital implementations is described. Main emphasis is on the presentation of the SP-300 flight-control system for the Boeing 737-300 aircraft. This dual diverse processor includes software diversity.

202. N. C. J. Wright, "Dissimilar Software," in *Workshop 'Design Diversity in Action'*, Baden, Austria, 27-28 June 1986.

203. S. Yoshimura, "Strategy for Back-to-Back Testing in the Project on Diverse Software (PODS)," HWR-97, OECD Halden Reactor Project, May 1983.

The Back-to-Back testing is one of the testing phases besides local testing and acceptance testing applied in the PODS experiment. The method of test data selection is described, and some test cases are presented. (Compare contribution by Bishop in this book.)

204. L. J. Yount, "Architectural Solutions to Safety Problems of Digital Flight-Critical Systems for Commercial Transports," in *Proc. of the AIAA/IEEE 6th Digital Avionics Systems Conf.*, pp. 28-35, Baltimore, MD, USA, 3-6 December 1984.

Starting from a description of the dual diverse SP-300 flight control system, new architectures for future applications are presented. Techniques to overcome generic hardware and software faults are discussed.

205. L. J. Yount, K. A. Liebel, and B. H. Hill, "Fault Effect Protection and Partitioning for Fly-by-Wire and Fly-by-Light Avionics Systems," in *Proc. AIAA/ACM/NASA/IEEE Computers in Aerospace V Conference*, pp. 275-284, Long Beach, CA, USA, 21-23 October 1985.

The historical development of automatic landing systems from analog implementations to current digital ones is discussed. The techniques used to tolerate generic software faults are described, especially the use of multi version software. The SP-300 autopilot flight director system used in B737-300 aircraft and a new dual/dual fault tolerant architecture are presented.

206. L. J. Yount, "Generic Fault-Tolerance Techniques for Critical Avionics Systems," in *Proc. AIAA Guidance and Control Conference*, Snowmass, CO, USA, June 1985.

Different categories of generic faults, which are design faults in hardware and in software, are discussed, and the benefits from applying software fault tolerance techniques are explained. A quantification of the reliability gain of a three version system is given, which makes the ideal (and unrealistic) assumption of total independence.

207. L. J. Yount, "Use of Diversity in Boeing Airplanes," in *Workshop 'Design Diversity in Action'*, Baden, Austria, 27-28 June 1986.

208. A. Zeh, "Softwareentwicklung für ein zuverlässiges und sicheres Prozeßrechnersystem (Software Development for a Reliable and Safe Process Control System - in German)," in *Proc. Fachtagung Prozeßrechner 1981*, pp. 240-250, München, Germany, 10-11 March 1981.

Design of a railway safety system with dual diverse hardware and dual diverse software. The concept was not implemented.